Telecommunication
Transmission Principles

Telecommunication Transmission Principles

Edited by
Fraidoon Mazda
MPhil DFH CEng FIEE

With specialist contributions

Focal Press
An imprint of Butterworth-Heinemann
Linacre House, Jordan Hill, Oxford OX2 8DP
A division of Reed Educational and Professional Publishing Ltd

$\mathcal{R}$ A member of the Reed Elsevier plc group

OXFORD BOSTON JOHANNESBURG
NEW DELHI SINGAPORE MELBOURNE

First published 1996

British Library Cataloguing in Publication Data
Mazda, Fraidoon F
    Telecommunication Transmission Principles
    I. Title
    621.382

ISBN 02405 1452 1

Library of Congress Cataloguing in Publication
Mazda, Fraidoon F.
    Telecommunication Transmission Principles/Fraidoon Mazda
    p. cm.
    Includes bibliographical references and index.
    ISBN 02405 1452 1
    1. Telecommunications. I. Title
    TK5101.M37 1993   92-27846
    621.382–dc20   CIP

Printed and bound in Great Britain by
Biddles Ltd, Guildford and King's Lynn

Contents

Preface

The transmission of information is one of the basic requirements of telecommunication systems, and this book reviews the principles behind the techniques and components used in this transmission process.

Modulation makes the signal suitable for transmission over a transmission medium, whether it is copper, fibre or the air waves. Analogue modulation is described in Chapter 1 and digital modulation in Chapter 2. Multiplexing of several signals over a common carrier is also an essential part of telecommunications transmission. Chapter 3 describes frequency modulation and Chapter 4 covers the more common time division multiplexing.

The general principles of digital transmission are described in Chapter 5. These includes: design considerations, such as noise; transmission errors and their use in performance measures; line codes; clocking considerations, including jitter; and framing.

Chapters 6 and 7 describe two of the prime physical components used for transmission, i.e. copper and fibre optic systems. Wireless communication is described in a separate book in this series. The construction and characteristics of copper and fibre optic cables are covered in Chapter 6 and Chapter 7 describes fibre optic components, such as sources, detectors and systems, in greater detail.

Nine authors have contributed to this book, all specialists in their field, and the success of the book is largely due to their efforts. The book is also based on selected chapters which were first published in the much larger volume of the *Telecommunications Engineers' Reference Book*, now in its sixth edition.

Fraidoon Mazda
Bishop's Stortford
April 1996

List of contributors

Luc Ceuppens
Ind Ing
Telindus Ltd
(Chapter 4)

Professor J E Flood
OBE DSc FInstP CEng FIEE
Aston University
(Chapters 1 and 2)

Takis Hadjifotiou
Bell Northern Research
(Chapter 7, Sections 7.1–7.6)

P J Howard
Telecoms Consultant
(Chapter 6)

Edwin V Jones
BSc MSc PhD CEng MIEE
University of Essex
(Chapter 5)

David Lockstone
Bell Northern Research
(Chapter 3)

Fraidoon Mazda
MPhil DFH CEng FIEE
Bell Northern Research
(Chapter 8)

John McFarlane
Nortel Ltd.
(Chapter 7, Section 7.7)

Michael J Simmons
Telindus Ltd.
(Chapter 4)

1. Analogue modulation

1.1 Introduction

The processing of a signal to make it suitable for sending over a transmission medium is called modulation.

Reasons for using modulation are:

1. Frequency translating (e.g. when an audio frequency baseband signal modulates a radio frequency carrier).
2. Improving signal/noise ratio by increasing the bandwidth (e.g. using frequency modulation).
3. Multiplexing i.e. enabling many baseband channels to share the same wideband transmission path.

Modulation is performed by causing the baseband modulating signal to vary a parameter of a carrier wave. A sinusoidal carrier, as given by Equation 1.1, is defined by three parameters: Amplitude A, Frequency $\omega / 2\pi$ and Phase φ.

Thus there are three basic modulation methods:

1. Amplitude modulation (AM).
2. Frequency modulation (FM).
3. Phase modulation (PM).

$$v_c = A \cos(\omega t + \varphi) \tag{1.1}$$

When modulation is employed, a modulator is needed at the sending end of a channel and a demodulator at the receiving end recovers the baseband signal from the modulated carrier. The combination of modulator and demodulator at a terminal is often referred to as a modem.

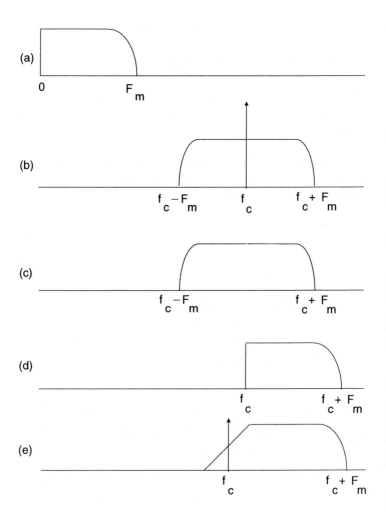

Figure 1.1 Frequency spectra for amplitude modulation:
(a) baseband signal; (b) simple amplitude modulation (AM);
(c) double sideband suppressed carrier (DSBSC) modulation;
(d) single sideband suppressed carrier (SSBSC) modulation;
(e) vestigial sideband (VSB) modulation

1.2 Amplitude modulation

1.2.1 Simple amplitude modulation

The simplest form of modulation is amplitude modulation. The modulator causes the envelope of the carrier wave to follow the waveform of the modulating signal and the demodulator recovers it from this envelope.

If a carrier, given by Equation 1.2, is modulated to a depth m by a sinusoidal modulating signal given by Equation 1.3, the resulting AM signal is as in Equation 1.4.

$$v_c = V_c \cos \omega_c t \qquad (1.2)$$

$$v_m = V_m \cos \omega_m t \qquad (1.3)$$

$$\begin{aligned} v &= (1 + m \cos \omega_m t) V_c \cos \omega_c t \\ &= V_c [\cos \omega_c t + \frac{1}{2} m \cos(\omega_c + \omega_m) t \\ &\quad + \frac{1}{2} m \cos(\omega_c - \omega_m) t] \end{aligned} \qquad (1.4)$$

If the modulating signal contains several components, f_1, f_2, ..., etc., then the modulated signal contains f_c-f_1, f_c-f_2, ..., etc., and f_c+f_1, f_c+f_2, ..., etc. in addition to f_c. If the modulating signal consists of a band of frequencies, as shown in Figure 1.1(a), the modulated signal consists of two sidebands, each occupying the same bandwidth as the baseband signal, as shown in Figure 1.1(b). In the upper sideband, the highest frequency corresponds to the highest frequency in the baseband; this is therefore known as an erect sideband. In the lower sideband, the highest frequency corresponds to the lowest frequency in the baseband; this is known as an inverted sideband.

Simple amplitude modulation makes inefficient use of the transmitted power, as information is transmitted only in the sidebands but the majority of the power is contained in the carrier. If a carrier as in

Equation 1.2 is modulated to a depth m by the sinusoidal baseband signal of Equation 1.3, the output power is given by Equation 1.5.

$$\overline{v^2} = V_c^2 \left(\frac{1}{2} + \frac{1}{8} m^2 + \frac{1}{8} m^2 \right)$$
$$= \frac{1}{2} V_c^2 \left(1 + \frac{1}{2} m^2 \right) \tag{1.5}$$

The maximum sideband power is obtained with 100% depth of modulation. The power in the sidebands is then a third of the total transmitted power. For smaller modulation depths, it is even less.

One method of producing AM is to add the baseband signal to the carrier and apply them to a non-linear amplifier, as shown in Figure 1.2. If the input/output characteristic of the non-linear circuit is given by Equation 1.6 and its input voltage by Equation 1.7, then the output voltage is as in Equation 1.8.

$$v_o = a_o + a_1 v_i + a_2 v_i^2 + \ldots \tag{1.6}$$

$$v_i = v_m \cos \omega_m t + V_c \cos \omega_c t \tag{1.7}$$

$$\begin{aligned}
v_o = {} & a_o + a_2 (V_m^2 + V_c^2) \\
& + a_1 (V_m \cos \omega_m t + V_c \cos \omega_c t) \\
& + a_2 (\frac{1}{2} V_m^2 \cos 2\omega_m t + \frac{1}{2} V_c^2 \cos 2\omega_c t) \\
& + V_c V_m [\cos(\omega_c - \omega_m) t \\
& + \cos(\omega_c + \omega_m) t]) + \ldots
\end{aligned} \tag{1.8}$$

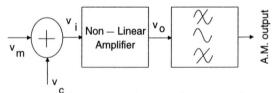

Figure 1.2 Low level amplitude modulator

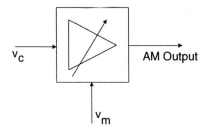

Figure 1.3   High level amplitude modulator

The bandpass filter is required to remove all components except those which comprise the AM wave. These are shown in bold in the above equation.

Another form of modulator is shown in Figure 1.3. This uses a gain controlled amplifier whose input signal is the carrier and whose control voltage is the modulating signal. For example, the carrier may be applied to the base of a transistor and the modulating signal superimposed on the collector supply voltage.

If the gain of the amplifier is given by Equation 1.9 and the carrier and modulating waveforms by Equations 1.2 and 1.3 then the output voltage is given by Equation 1.10, which is the required AM wave.

$$A = A_o (1 + k v_m) \tag{1.9}$$

$$\begin{aligned}
v_o &= A_o V_c \cos \omega_c t + k A_o V_m V_c \cos \omega_m t \cos \omega_c t \\
&= A_o V_c (\cos \omega_c t + \frac{1}{2} k V_m [\cos (\omega_o - \omega_m) t \\
&+ \cos (\omega_c + \omega_m) t])
\end{aligned} \tag{1.10}$$

The first method is used for low level modulators before the power amplifier of a transmitter. The second method is used for high level modulators in the final stage of amplification.

An AM wave can be demodulated by the simple diode circuit shown in Figure 1.4. The rectified output voltage across the load resistor follows the envelope of the modulated input signal as in Figure 1.5. The time constant CR must be large compared with the

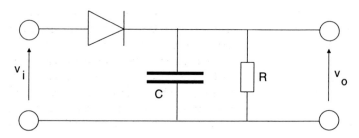

Figure 1.4 Envelope demodulator circuit

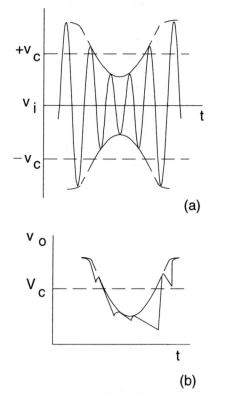

Figure 1.5 Envelope demodulator waveforms: (a) input; (b) output

period of the carrier to prevent the output voltage decaying substantially between the peaks of the carrier. However, time constant CR in Figure 1.4 must be sufficiently small for the output voltage to decay as rapidly as the envelope changes when the baseband signal has its maximum frequency, F_m. If $f_c >> F_m$, this is easily arranged.

Other demodulators, which give a better performance at poor input signal/noise ratios, are the coherent demodulator (described later) and the phase locked loop AM demodulator (Gardner, 1979; Gosling, 1986).

1.2.2 Suppressed carrier modulation

It is possible, by using a balanced modulator (Tucker, 1953) to eliminate the carrier and generate only the sidebands, as shown in Figure 1.1(c). This is known as double sideband suppressed carrier modulation (DSBSC). The modulator acts as a switch, which multiplies the baseband signal by a quasi square wave carrier. Its output thus contains upper and lower sidebands about the fundamental and harmonics of the carrier frequency. As shown in Figure 1.6(a), a bandpass filter is used to remove all components except the wanted sidebands. Thus, for a sinusoidal baseband input signal, the output signal is given by Equation 1.11.

$$v_o = V_m \cos \omega_m t \cos \omega_c t$$

$$= \frac{1}{2} V_m \left(\cos (\omega_c - \omega_m) t + \cos (\omega_c + \omega_m) t \right) \quad (1.11)$$

A balanced modulator used in this way is often called a product modulator.

To demodulate a DSBSC signal, it is necessary to use a coherent demodulator, consisting of a balanced modulator supplied with a locally generated carrier as shown in Figure 1.6(b) instead of the envelope demodulator used with simple AM. If the incoming DSBSC signal is given by Equation 1.12 and the coherent demodulator multiplies this with a local carrier given by Equation 1.13, its output voltage is as in Equation 1.14.

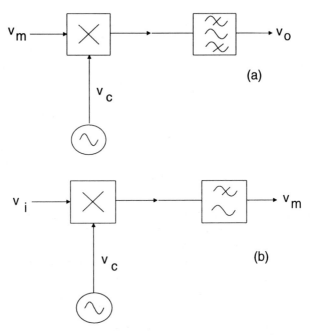

Figure 1.6 Application of balanced modulator: (a) product modulator; (b) coherent demodulator

$$v_i = \frac{1}{2} m V_c \left(\cos (\omega_c + \omega_m) t + \cos (\omega_c - \omega_m) t \right) \qquad (1.12)$$

$$v_c = \cos (\omega_c t + \theta) \qquad (1.13)$$

$$
\begin{aligned}
v = \frac{1}{4} m V_c (&\cos [(2 \omega_c + \omega_m) t + \theta] \\
&+ \cos [(2 \omega_c - \omega_m) t + \theta] \\
&+ \cos (\theta + \omega_m t) + \cos (\theta - \omega_m t))
\end{aligned} \qquad (1.14)
$$

The components at frequencies ($2\omega_c \pm \omega_m$) are removed by a low pass filter and the baseband output signal is given by Equation 1.15.

$$v_o = \frac{1}{4} m V_c \left(\cos(\theta + \omega_m t) + \cos(\theta - \omega_m t) \right)$$

$$= \frac{1}{2} m V_c \cos\theta \cos\omega_m t \qquad (1.15)$$

Thus v_o represents the original baseband signal, provided that the phase θ of the local carrier is stable.

A further economy in power, and a halving in bandwidth, can be obtained by producing a single sideband suppressed carrier (SSBSC) signal, as shown in Figure 1.1(d). If the upper sideband is used, the effect of the modulator is simply to produce a frequency translation of the baseband signal to a position in the frequency spectrum determined by the carrier frequency. If the lower sideband is used, the band is inverted as well as translated.

The SSBSC signal requires the minimum possible bandwidth for transmission. Consequently, the method is used whenever its complexity is justified by the saving in bandwidth (Pappenfus et al., 1964). An important example is the use of SSBSC for multichannel carrier telephone systems (Kingdom, 1991).

An error in the frequency of the local carrier of the demodulator results in a corresponding shift in the frequencies of the components in the baseband output signal. For speech transmission, frequency shifts of the order of ± 10Hz are not noticeable, but the errors that can be tolerated for telegraph and data transmission are less. The ITU-T (formerly CCITT) specifies that the frequency shift should be less than ± 2Hz.

A SSBSC signal can be generated by using a balanced modulator and a bandpass filter, as for DSBSC. However, the filter in Figure 1.6(a) is designed to pass only one of the sidebands, instead of both.

An alternative method of generating a SSBSC signal is the quadrature method shown in Figure 1.7. This uses two product modulators. The baseband signal, v_m, and the carrier, v_c, are applied to one directly, and to the other after a phase shift of 90°. If the carrier and modulating signals are as in Equations 1.2 and 1.3, the outputs of the two modulators are given by Equations 1.16 and 1.17 and v_o by Equation 1.1.

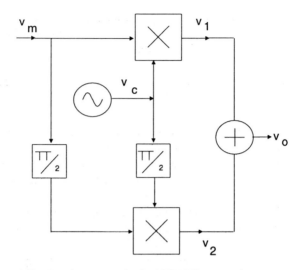

Figure 1.7 Quadrature method of SSBSC generation

$$v_1 = V_m \sin \omega_m t \sin \omega_c t \tag{1.16}$$

$$v_2 = V_m \cos \omega_m t \cos \omega_c t \tag{1.17}$$

$$v_o = v_1 + v_2 = V_m \cos(\omega_c - \omega_m)t \tag{1.18}$$

Alternatively, subtracting v_1 from v_2 gives Equation 1.19.

$$v_o = V_m \cos(\omega_c + \omega_m)t \tag{1.19}$$

It is straightforward to provide a 90° phase shift at the single carrier frequency, but not over the band of frequencies of a modulating signal. Instead, the baseband signal is applied to the modulators through a pair of networks whose phase shifts differ by approximately 90° over the required band (Coates, 1975). The quadrature method is not widely used, since small phase errors result in unwanted frequency components in the output signal.

A coherent demodulator is required for demodulating a SSBSC signal. Since only one sideband is present, the demodulated output signal, from Equation 1.15, is as in Equation 1.20.

$$v_o = \frac{1}{4} m V_c \cos (\omega_m t \pm \theta)$$

(1.20)

If a SSBSC signal is corrupted by noise, the coherent demodulator acts on each component of the noise spectrum in the same way as it does on each component of the signal. Consequently, the signal to noise ratio at the output of the demodulator is the same as the input signal to noise ratio.

For a DSBSC signal, the output of the demodulator produced by a signal component at $f_c + f_m$ is in phase with that produced by the component at $f_c - f_m$. Thus, the output signal voltage is twice the input signal voltage (c.f. Equation 1.15 with $\theta = 0$) and the output signal power is four times that for SSBSC. For noise, the outputs produced by components at $f_c - f_m$ and $f_c + f_m$ have a random phase difference. Thus, the output noise power is twice the input noise power. Consequently, the output signal to noise ratio is twice the input signal to noise ratio, an apparent improvement of 3dB over SSBSC. However, the bandwidth required for DSBSC transmission is twice that for SSBSC. This doubles the received noise power, so no overall improvement is obtained.

For simple AM, a coherent demodulator acts in the same way as for DSBSC. So does the envelope demodulator of Figure 1.4 at good signal to noise ratios. The diode acts as a switch operated by the incoming carrier and this multiplies the input signal by $\cos \omega_c t$, as does a coherent demodulator.

At poor input signal to noise ratios, this is no longer true. The diode is switched by peaks of the noise voltage and the output signal to noise ratio is worse than for a coherent demodulator (Brown and Glazier, 1974; Coates, 1975). The signal to noise ratio performance of simple AM is, of course, worse than that of DSBSC because the input power includes that of the carrier as well as the sidebands. For 100% depth of modulation, the sideband power is only a third of the total power. Thus, for equal transmitter powers and receiver noise power densities, simple AM gives an output signal to noise ratio 4.8dB worse than for DSBSC or SSBSC transmission.

By using SSBSC, it is possible to transmit two channels through the bandwidth needed by simple AM for a single channel; one uses the upper sideband of the carrier and the other uses the lower sideband. This is known as independent sideband modulation and is used in h.f. radio communication (Hills, 1973). However, it is also possible to do this using DSBSC. The transmitter uses two modulators whose carriers are in quadrature. The receiver uses two coherent demodulators whose local carriers are in quadrature. Equation 1.15 shows that each demodulator produces a full output from the signal whose carrier is in phase (since cos 0 = 1) and zero output from the signal whose carrier is in quadrature (since cos π/2 = 0). This is called quadrature amplitude modulation (QAM). In practice, the method is not used for analogue baseband signals since small errors in the phases of the local carriers cause a fraction of the signal of each channel to appear as crosstalk in the output from the other. However, the method is used for transmitting digital signals.

1.2.3 Vestigial sideband modulation

If the sideband signal extends down to very low frequencies, as in television, it is almost impossible to suppress the whole of the unwanted sideband without affecting low frequency components in the wanted sideband. Use is then made of vestigial sideband (VSB) transmission instead of SSBSC. A conventional AM signal (as shown in Figure 1.1(b)) is first generated and this is then applied to a filter having a transition between its pass and stop band that is skew symmetric about the carrier frequency. This results in an output signal having the spectrum shown in Figure 1.1(e).

If a coherent demodulator is used, the original baseband signal can be recovered without distortion. It is also possible to use a simple envelope demodulator for VSB, but some non linear distortion then results (Black, 1953). VSB transmission does, of course, require a greater channel bandwidth than SSB. However, for a wideband signal such as television, the bandwidth saving compared with DSB is considerable.

1.3 Frequency and phase modulation

1.3.1 General

The instantaneous angular frequency of an alternating voltage is given by Equation 1.21. This relationship between frequency and phase means that frequency modulation (FM) and phase modulation (PM) are both forms of angle modulation.

$$\omega = \frac{d\varphi}{dt} \quad radians/s \tag{1.21}$$

A sinusoidal carrier modulated by a sinusoidal baseband signal may be represented by Equation 1.22 where β is the modulation index.

$$v = V_c \cos\left(\omega_c t + \beta \sin \omega_m t\right) \tag{1.22}$$

The maximum phase deviation is given by Equation 1.23 and, since Equation 1.24 holds, the maximum frequency deviation is given by Equation 1.25.

$$\Delta\varphi = \pm\beta \tag{1.23}$$

$$\frac{d\varphi}{dt} = \omega_m \beta \cos \omega_m t \tag{1.24}$$

$$\Delta F = \pm\beta f_m \tag{1.25}$$

In PM the phase deviation is proportional to the modulating voltage; therefore β is independent of its frequency and the frequency deviation is proportional to it. In FM, the frequency deviation is proportional to the modulating voltage; therefore the deviation frequency is independent of its frequency and β is inversely proportional to it. Thus, for FM, the modulation index may be defined as in Equation 1.26.

$$\beta = \frac{maximum\ frequency\ deviation\ of\ carrier}{maximum\ baseband\ frequency}$$

$$= \frac{\Delta F}{F_m} \tag{1.26}$$

In angle modulation, the information is conveyed by the instantaneous phase of the signal. Consequently, phase distortion in the transmission path causes attenuation distortion of the received signal. The differential delay of the transmission path must therefore be closely controlled over the bandwidth required to transmit the signal. However, since the information is not conveyed by the amplitude of the signal, the receiver can contain a limiter to maintain a constant signal amplitude.

Consequently, non linear distortion in the transmission path does not cause distortion of the demodulated output signal; nor does attenuation/frequency distortion.

Use of a limiter in the receiver also removes amplitude variations due to noise and so enables angle modulation to give a better output signal to noise ratio than AM.

Moreover, since the amplitude of an angle modulated wave is constant, the transmitter can deliver its full rated power all the time, whereas for AM this only occurs for peak amplitudes of the baseband signal. This contributes a further improvement in signal to noise ratio.

In PM, the frequency deviation (βf_m) is proportional to the frequency of the modulating signal as well as to its amplitude.

Consequently, for signals (such as speech) that have the major proportion of their energy at the lower end of the baseband, PM makes inefficient use of the transmission path bandwidth compared with FM.

Moreover, to demodulate a PM signal, the receiver must compare the phase of the incoming carrier with that of a locally generated carrier which must be very stable.

FM is therefore preferred to PM for the transmission of analogue signals. It is used whenever sufficient bandwidth can be provided. However, PM is widely used for the transmission of digital signals as described in Chapter 2.

1.3.2 Modulators and demodulators

To generate FM, a voltage controlled oscillator is needed to enable the instantaneous frequency of the carrier to be varied by the baseband signal. This usually employs a varactor diode or a transistor reactance circuit to provide a voltage controlled capacitance (Coates, 1975).

The frequency of the oscillator is given by Equation 1.27. Thus, if C is given by Equation 1.28 then Equation 1.29 may be obtained.

$$f = \frac{1}{2 \pi \sqrt{L\,C}} \tag{1.27}$$

$$C = C_o\,(\,1\,+\,k\,v_m\,) \tag{1.28}$$

$$f = \frac{1}{2 \pi \sqrt{L\,C_o}} \left(\,1 - \frac{1}{2}\,k\,v_m + \frac{3}{8}\,k^2\,v_m^2 + ...\,\right) \tag{1.29}$$

To achieve linearity, it is therefore necessary for the frequency deviation to be kept to a small fraction of the centre frequency. Since the frequency depends on device parameters, which may change, a frequency stabilisation loop is usually added to control the centre frequency from a crystal oscillator.

Phase modulation can be generated by Armstrong's method shown in Figure 1.8. A product modulator produces a DSBSC wave, as shown in Figure 1.9(a), and the carrier is added to this after shifting its phase by 90°. This produces a phase deviation of the resultant signal, V_o, as shown in Figure 1.9(b).

This method is only suitable for small values of modulation index. It produces an output containing only two side frequencies, whereas it is shown in the next section that significant additional side frequencies are present with high index modulation. These higher order side frequencies serve to keep the amplitude of the signal constant, whereas the circuit of Figure 1.8 produces some residual amplitude modulation.

The locus of the end of phasor V_o is a straight line, whereas it would be a circle for perfect phase modulation.

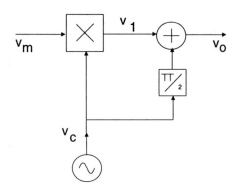

Figure 1.8   Phase modulation by Armstrong's method

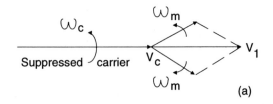

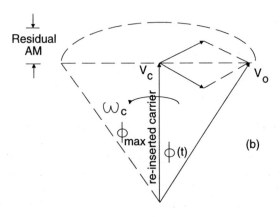

Figure 1.9   Phasor diagrams for Armstrong's phase modulator of Figure 1.8: product modulator output (V_1): output from circuit (V_0)

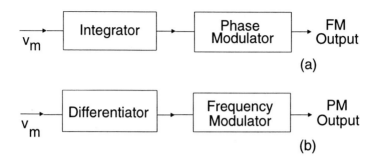

Figure 1.10   Alternative methods of generating angle modulation: (a) use of phase modulator to generate FM; (b) use of frequency modulator to generate PM

Frequency modulation may be produced by using a phase modulator whose modulating signal is the integral of the baseband signal, as shown in Figure 1.10. Therefore Equations 1.30 and 1.31 may be obtained and this corresponds to FM.

$$v_o = V_c \cos\left(\omega_c t + k \int_o^t v_m \, dt \right)$$

$$= V_c \cos\left(\omega_c t + \varphi(t) \right) \tag{1.30}$$

$$\omega = \omega_c + \frac{d\varphi}{dt}$$

$$= \omega_c + k v_m \tag{1.31}$$

The Armstrong modulator of Figure 1.8 can therefore be used to generate FM. Similarly, phase modulation may be generated with a frequency modulator by differentiating the baseband signal, as shown in Figure 1.10(b). The Armstrong circuit only produces low index modulation. The frequency deviation can be increased by applying its output to a non-linear circuit to generate harmonics, one of which is selected by a bandpass filter. If the nth harmonic is selected, the frequency deviation is increased by a factor n.

A phase modulated wave can be demodulated by using a coherent demodulator whose local carrier is in quadrature with the incoming carrier when unmodulated. When the incoming carrier is modulated, the output voltage is given by Equation 1.32.

$$v = V_c \cos \left(\omega_c t + \varphi(t) \right) \sin \omega_c t$$

$$= \frac{1}{2} V_c \cos \left(2 \omega_c t + \varphi(t) \right) + \frac{1}{2} V_c \sin \varphi(t) \tag{1.32}$$

The component at frequency $2 \pi \omega_c$ is removed by the low pass filter and the output signal is as in Equation 1.33.

$$v_o = \frac{1}{2} V_c \sin \varphi(t)$$

$$\approx \frac{1}{2} V_c \varphi(t) = k v_m \tag{1.33}$$

One method of demodulating FM is to use the incoming carrier to generate a train of pulses and to count the number of these per unit of time (Gosling, 1986). However, more widely used circuits operate indirectly by converting frequency variations of the incoming signal to variations of amplitude or phase and demodulating these. Such circuits are called discriminators.

A simple discriminator can use a parallel tuned circuit whose resonant frequency is offset from the centre frequency of the incoming FM signal. As shown in Figure 1.11(a), a deviation of the input frequency varies the output voltage and this is rectified by an envelope demodulator to give a baseband output voltage. Improved forms of discriminator use a pair of tuned circuits, one resonant on each side of the centre frequency, to obtain linearity of conversion over a greater range of frequency deviation (Langford-Smith, 1960; Sturley, 1965). Since the output voltage of a discriminator is proportional to the amplitude of the input signal as well as its frequency, the discriminator must be preceded by a limiter to ensure that its input is of constant amplitude.

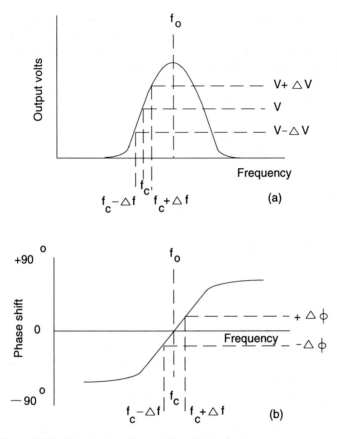

Figure 1.11   Principles of tuned circuit discriminators:
(a) frequency to amplitude conversion; (b) frequency to phase
conversion

A tuned circuit can also produce a phase shift depending on the
frequency deviation, as shown in Figure 1.11(b). The baseband signal
can therefore be recovered by a coherent demodulator. The incoming
FM carrier is fed directly to the coherent demodulator, but it is shifted
90° (by a small series capacitance) before being applied to the
discriminator. The circuit is therefore known as a quadrature FM
demodulator (Gosling, 1986).

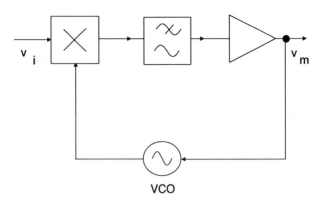

Figure 1.12   Phase locked loop circuit for demodulating FM

A form of demodulator which is increasingly used for FM is the phase locked loop circuit shown in Figure 1.12. When the voltage controlled oscillator (VCO) has zero control voltage, its output frequency is that of the unmodulated carrier. When the frequency of the input signal begins to change, the resulting phase difference between it and the output of the VCO causes the coherent demodulator to produce an output voltage. This is fed back to the VCO to change its frequency and so reduce the phase error. In this way, the frequency of the VCO tracks the varying frequency of the incoming carrier. If the characteristic of the VCO is linear, its control voltage is thus proportional to the incoming carrier frequency and provides a demodulated baseband output signal. The low pass filter eliminates noise and interference at frequencies above the baseband.

The phase locked loop demodulator is less complicated than a discriminator because it has no tuned circuits to align. It also has a better output signal to noise performance at low values of input signal to noise ratio (Pappenfus et al., 1964; Viterbi, 1966; Roberts, 1977).

1.3.3   Frequency spectra

From Equation 1.22, a sinusoidal carrier modulated by a sinusoidal baseband signal may be represented by Equation 1.34.

$$v = R \, V_c \, e^{j \, \omega_c \, t} \, e^{j \, \beta \, \sin \omega_m \, t} \tag{1.34}$$

However Equation 1.35 holds, where $J_n(\beta)$ is the Bessel function of the first kind of order n, resulting in Equation 1.36.

$$e^{j \, \beta \, \sin \theta} = \sum_{n=-\infty}^{\infty} J_n(\beta) \, e^{j \, n \, \theta} \tag{1.35}$$

$$v = R \, V_c \, e^{j \, \omega_c \, t} \sum_{n=-\infty}^{\infty} J_n(\beta) \, e^{j \, n \, \omega_m \, t}$$

$$= V_c \sum_{n=-\infty}^{\infty} J_n(\beta) \cos(\omega_c + n \, \omega_m) \, t \tag{1.36}$$

Thus, the spectrum of the angle modulated wave has an upper and a lower sideband, each containing frequencies separated from the carrier by all the harmonics of the baseband signal.

Equation 1.36 shows that angle modulation is non linear. If the baseband signal contains two frequencies, f_1 and f_2, the sidebands contain many components at frequencies $f_c \pm qf_1 \pm rf_2$ (where q = 1,2,3,..., r = 1,2,3,...) in addition to $f_c \pm nf_1$ and $f_c \pm nf_2$. If the spectrum of the baseband signal contains many components, the spectrum of each sideband is extremely complicated (Black, 1953). It can be represented by a band of Gaussian noise (Roberts, 1977).

Tables of $J_n(x)$ have been published (Jahnke and Emde, 1945). Some values of $J_n(\beta)$ are given in Table 1.1 and some typical voltage spectra of angle modulated signals are shown in Figure 1.13. It will be seen that for low index modulation ($\beta < 1$) only the components at $f_c \pm f_m$ are significant. Thus, the bandwidth required for transmission is no greater than for AM. However, for high index modulation, the bandwidth is much greater than for AM.

In general, the magnitudes of $J_n(\beta)$ are small for $n > \beta + 1$. As a result, the bandwidth W required is approximately given by Equation 1.37 (Carson's rule).

Table 1.1 Bessel functions of the first kind

β	$J_0(\beta)$	$J_1(\beta)$	$J_2(\beta)$	$J_3(\beta)$	$J_4(\beta)$	$J_5(\beta)$	$J_6(\beta)$	$J_7(\beta)$	$J_8(\beta)$	$J_9(\beta)$	$J_{10}(\beta)$
0	1.000										
0.2	0.990	0.099	0.005								
0.4	0.960	0.196	0.019	0.001							
0.6	0.912	0.286	0.043	0.004							
0.8	0.846	0.368	0.075	0.010	0.001						
1.0	0.765	0.440	0.114	0.019	0.002						
2.0	0.223	0.576	0.352	0.128	0.034	0.007	0.001				
3.0	−0.260	0.339	0.486	0.309	0.132	0.043	0.011	0.002			
4.0	−0.397	−0.066	0.364	0.430	0.281	0.132	0.049	0.015	0.004		
5.0	−0.177	−0.327	0.046	0.364	0.391	0.261	0.131	0.053	0.018	0.005	0.001
6.0	0.150	−0.276	−0.242	0.114	0.357	0.362	0.245	0.129	0.056	0.021	0.006
7.0	0.300	−0.004	−0.301	−0.167	0.157	0.347	0.339	0.233	0.128	0.058	0.023
8.0	0.171	0.234	−0.113	−0.291	−0.105	0.185	0.337	0.320	0.223	0.126	0.060
9.0	−0.090	0.245	0.144	−0.180	−0.265	−0.055	0.204	0.327	0.305	0.214	0.124
10.0	−0.245	0.045	0.254	0.058	−0.219	−0.234	−0.014	0.216	0.317	0.291	0.207

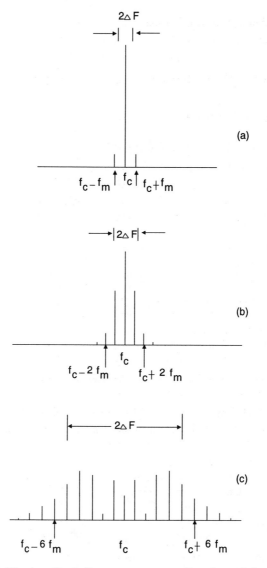

Figure 1.13 Amplitude/frequency spectra of angle modulated waves: (a) $\beta = 0.2$; (b) $\beta = 1.0$; (c) $\beta = 5.0$

$$W = 2 F_m (1 + \beta) \tag{1.37}$$

Thus, for FM, Equation 1.38 may be obtained where ΔF is the maximum frequency deviation.

$$W = 2 (F_m + \Delta F) \tag{1.38}$$

1.3.4 Interference and noise

In FM, the amplitude of the signal conveys no information. The receiver contains a limiter, so amplitude variations due to interference and noise (unless very severe) have no effect. However, interference and noise also perturb the phase of the signal and thus its instantaneous frequency.

The phasor diagram in Figure 1.14(a) represents an unmodulated carrier with a small interfering signal separated from it by a frequency difference f_i. If the interfering voltage $V_x \cos (\omega_c + \omega_i)t$ is added to the carrier $V_c \cos \omega_c t$, the resultant is as in Equation 1.39.

$$v = V_c \cos \omega_c t + V_x \cos (\omega_c + \omega_i) t$$

$$= (V_c + V_x \cos \omega_i t) \cos \omega_c t - V_x \sin \omega_c t \sin \omega_i t \tag{1.39}$$

The first term represents amplitude modulation of the carrier and the second term represents a phase displacement, $\varphi_i (t)$, given by Equation 1.40.

$$\varphi_i (t) = \tan^{-1} \left(\frac{V_x \sin \omega_i t}{V_c + V_x \cos \omega_i t} \right) \tag{1.40}$$

But $V_x \ll V_c$, so Equation 1.41 may be obtained.

$$\varphi_i (t) \approx \frac{V_x}{V_c} \sin \omega_i t \tag{1.41}$$

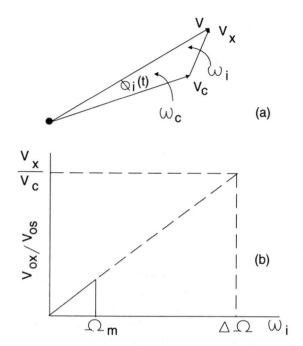

Figure 1.14   Effect of interfering signal on FM receiver:
(a) phasor diagram of input signal; (b) output signal due to
interference

For FM, the demodulated output voltage is proportional to $d\varphi/dt$.
Thus, the output voltage due to the interference is as in Equation 1.42,
for values of f_i within the pass band of the receiver's output low pass
filter, which corresponds to the baseband (F_m) to be handled. If $f_i >
F_m$, no interference reaches the output.

$$v_{ox} = k\frac{d\varphi}{dt} = \frac{k\,\omega_i\,V_x}{V_c}\cos\omega_i t \tag{1.42}$$

If the carrier is frequency modulated, the signal output voltage is
given by Equation 1.43, where $\Delta\Omega$ is the maximum deviation of
angular frequency.

$$v_{os} = k \, \Delta \Omega \cos \omega_m t = V_{os} \cos \omega_m t \qquad (1.43)$$

Thus, the ratio between the output interference and signal voltages is as in Equation 1.44.

$$\frac{V_{ox}}{V_{os}} = \frac{\omega_i V_i}{\Delta \Omega V_c} \qquad \text{for } \omega_i < \Omega_m$$

$$= 0 \qquad \text{for } \omega_i > \Omega_m \qquad (1.44)$$

For a large modulation index, $\Delta \Omega > \omega_m$. Thus, as shown in Figure 1.14(b), the output signal to interference ratio is much greater than V_c/V_x. It is therefore much better than that obtained with AM, the improvement being proportional to the modulation index $\Delta \Omega / \omega_m$.

If the interference is at the same frequency as the carrier (which is the worst situation for AM), Equation 1.44 shows that there is zero interference present at the receiver output.

Thus, when two transmissions at the same frequency are received, the output corresponds to the stronger. This is known as the capture effect.

Gaussian noise can be considered as consisting of an infinite number of interference components, $V_x(q)$ (where $q = 0, \pm 1, \pm 2, ...$), spaced at infinitesimal intervals $\delta \omega$. Since $V_{ox} \propto \omega_i$, the noise spectrum at the output of the receiver is parabolic, i.e. the noise power density increases with the square of the frequency.

The power of each component, $v_x(q)$, is equal to the noise power in its frequency band, giving Equation 1.45 where n is the input noise power density.

$$\frac{1}{2} V_x^2(q) = \overline{v_x^2(q)} = n \, \delta \omega \qquad (1.45)$$

The total output noise power, from Equation 1.42, is given by Equation 1.46.

$$N_o = \sum_{q=-\infty}^{\infty} k^2 \left(\frac{V_i(q)}{V_c} \right)^2 \omega_i^2(q)$$

$$= \frac{k^2 n}{V_c^2} \int_{-\Omega_M}^{\Omega_M} \omega^2 \, d\omega = \frac{2 k^2}{3 V_c^2} n \Omega_M^3 \qquad (1.46)$$

The output signal power, from Equation 1.43 is as in Equation 1.47.

$$S_o = \frac{1}{2} V_{os}^2 = \frac{1}{2} (k \Delta \Omega)^2 \qquad (1.47)$$

Thus, the output signal to noise ratio is given by Equation 1.48.

$$\frac{S_o}{N_o} = \frac{3 V_c^2 (\Delta \Omega)^2}{4 n \Omega_M^3} = \frac{3}{4} \frac{\beta^2 V_c^2}{n \Omega_m} \qquad (1.48)$$

The input signal power is as in Equation 1.49.

$$S_i = \frac{1}{2} V_c^2 \qquad (1.49)$$

If wideband (high index) FM is used, the bandwidth is approximately $2\Delta \Omega$, so the input noise power is given by Equation 1.50 and the input signal to noise ratio by Equation 1.51.

$$N_i = 2 n \Delta \Omega \qquad (1.50)$$

$$\frac{S_i}{N_i} = \frac{V_c^2}{4 n \Delta \Omega} \qquad (1.51)$$

Therefore Equation 1.52 may be obtained.

$$\frac{S_o/N_o}{S_i/N_i} = 3\,\beta^3 \tag{1.52}$$

Although the improvement in signal to noise ratio is proportional to β^3 (i.e. to ΔF^3), the input noise power is proportional to the bandwidth ($2\Delta F$). Thus, the output signal to noise ratio is proportional to the square of the bandwidth used. An octave increase in the frequency deviation and bandwidth increases the output signal to noise ratio by 6dB.

The bandwidth required is β times that for AM, so for equal noise power densities, the noise power at the input to the receiver is β times that for AM. Consequently, for fully modulated AM and the same carrier power, the improvement is given by Equation 1.53.

$$\tag{1.53}$$

$$\frac{\textit{Output signal noise ratio}\ \text{for}\ \textit{FM}}{\textit{Output signal noise ratio}\ \text{for}\ \textit{AM}} = 3\,\beta^2$$

For example, FM broadcasting typically has a baseband of 15kHz and a maximum frequency deviation of 75kHz, i.e. $\beta = 5$. The improvement in signal to noise ratio compared with AM with 100% modulation is $3 \times 25 = 75$, i.e. 19dB.

It has been shown above that the output noise spectrum for FM is parabolic. The output signal to noise ratio can therefore be improved further by making the signal power also increase with frequency. It is common practice to insert in front of the modulator at the transmitter a pre-emphasis network which has a rising gain-frequency characteristic across the baseband.

The receiver contains a de-emphasis network with the inverse gain-frequency characteristic, to obtain a channel having a flat overall gain-frequency characteristic and a uniform noise spectrum. For single channel telephony or broadcasting this provides a better output signal to noise ratio. When FM is used to transmit a wideband signal consisting of a block of telephone channels assembled by frequency division multiplexing, the use of pre-emphasis is essential. Otherwise, there would be a large difference between the signal to noise ratios of channels at the top and the bottom of the band.

It has been assumed above that amplitude variations due to noise have no effect on the output signal of a FM receiver. For this assumption to be valid, the input signal to noise ratio must exceed a threshold of approximately 12dB (Brown and Glazier, 1974). When the input signal to noise ratio decreases below this, peaks of noise begin to obliterate the carrier and the output signal to noise ratio deteriorates rapidly, as shown in Figure 1.15. Radio links used for commercial telecommunications must therefore have signal to noise ratios well above the threshold.

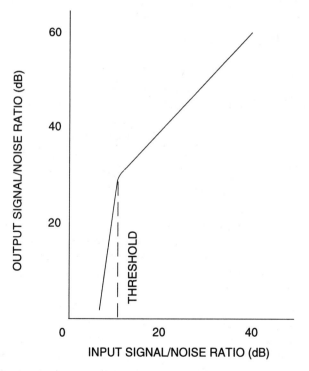

Figure 1.15   Improvement in signal to noise ratio given by frequency modulation ($\beta = 5.0$)

1.4 Pulse modulation

1.4.1 General

In the preceding sections, it has been assumed that the carrier wave is sinusoidal. However, it is also possible to modulate carriers having other waveforms. For example, it is possible to modulate trains of pulses to produce pulse amplitude modulation (PAM), pulse frequency modulation (PFM) or pulse phase modulation (PPM), which is sometimes called pulse position modulation. It is also possible to produce pulse length modulation, which is sometimes called pulse duration modulation (PDM) or pulse width modulation (PWM).

1.4.2 Pulse amplitude modulation

A basic PAM system is shown in Figure 1.16. If a train of pulses as shown in Figure 1.17(b) is amplitude modulated by the baseband signal shown in Figure 1.17(a) the resulting PAM signal is shown in Figure 1.17(c). The baseband signal is represented by a sequence of samples of it, so the process is also called sampling.

A train of pulses having a pulse repetition frequency (PRF) f_r may be represented by a Fourier series, as in Equation 1.54.

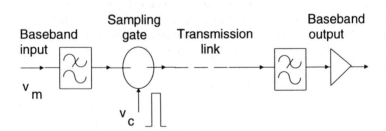

Figure 1.16 Basic pulse amplitude modulation system with one direction of transmission only

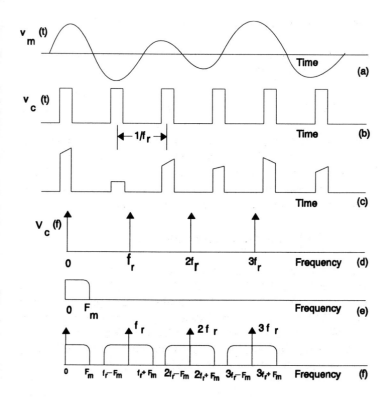

Figure 1.17   Pulse amplitude modulation waveforms:
(a) baseband signal; (b) unmodulated pulse train; (c) modulated
pulse train; (d) spectrum of unmodulated pulse train; (e) spectrum
of baseband signal; (f) spectrum of modulated pulse train

$$v_c = \frac{1}{2} a_o + \sum_{n=1}^{\infty} a_n \cos n\, \omega_r t \qquad (1.54)$$

Its spectrum thus contains a d.c. component, the PRF and its
harmonics, as shown in Figure 1.17(d). If the pulses are of short

duration, the amplitudes a_n of the harmonics are approximately equal up to large values of n.

If the pulse train is amplitude modulated by a sinusoidal baseband signal, the resulting PAM signal is given by Equation 1.55.

$$v = (1 + m \cos \omega_m t) v_c$$

$$= \frac{1}{2} a_o + \frac{1}{2} a_o\, m \cos \omega_m t + \sum_{n=1}^{\infty} a_n \cos n\, \omega_r t$$

$$= \frac{1}{2} m \sum_{n=1}^{\infty} a_n [\, (\cos (n\, \omega_r + \omega_r)\, t$$
$$+ \cos (n\, (\omega_r - \omega_m)\, t)\,] \tag{1.55}$$

The PAM signal contains all components of the original pulse, with the baseband frequency f_m and upper and lower side frequencies $(nf_r \pm f_m)$ about the PRF and its harmonics. If the modulating signal consists of a band of frequencies, (Figure 1.17(e)), the spectrum of the PAM signal contains the original baseband, with upper and lower sidebands about the PRF and its harmonics, (Figure 1.17(f)).

The PAM signal can be demodulated by means of a low pass filter which passes the baseband and stops the lower sideband of the PRF and all higher frequencies. For this to be possible, Equation 1.56 must hold, where F_m is the maximum frequency of the baseband signal and f_0 is the cut off frequency of the filter. Thus Equation 1.57 must hold.

$$F_m \leq f_o \leq f_r - F_m \tag{1.56}$$

$$f_r \geq 2 F_m \tag{1.57}$$

Equation 1.57 is a statement of the sampling theorem. This may be expressed as follows: if a signal is to be sampled and the original signal is to be recovered from the samples without error, the sampling frequency must be at least twice the highest frequency in the original signal. The sampling theorem is due to Nyquist and the lowest possible rate at which a signal may be sampled, $2F_m$, is often known as the Nyquist rate. If the sampling frequency is less than the Nyquist rate, the lower sideband of the PRF overlaps the baseband and it is impossible to separate them. The output from the low pass filter then contains unwanted frequency components; this situation is known as aliasing.

To prevent aliasing, it is essential to limit the bandwidth of the signal before sampling. Thus, practical systems pass the input signal through an anti-aliasing low pass filter of bandwidth $\frac{1}{2}f_r$ before sampling as shown in Figure 1.16. Practical filters are non-ideal; it is therefore necessary to have $f_r > 2F_m$ in order that the anti-aliasing filter and demodulating filter can have both very low attenuation at frequencies up to F_m and very high attenuation at frequencies down to $f_r - F_m$. For telephony, a baseband from 300Hz to 3.4kHz is provided and a sampling frequency of 8kHz is used. Thus $f_r - F_m = 4.6kHz$ and there is a guardband of 1.2kHz to accommodate the transition of the filters between their pass band and their stop band.

1.4.3 Pulse time modulation

Pulse frequency modulation, pulse phase modulation and pulse length modulation all vary the times of occurrence of the individual pulses of a pulse train. They can therefore be collectively called pulse time modulation. Since the modulating signal is not carried by the amplitude of the pulses, pulse time modulation can be almost immune to amplitude variations caused by interference and noise and so give a better output signal to noise ratio than PAM.

When pulse time modulation is used, the sidebands about the PRF and its harmonics contain sideband components at frequencies $nf_r \pm qf_m$ (where $q = 1,2,3,...$), just as do the sidebands of a sinusoidal carrier when angle modulation is used (Fitch, 1947; Moss, 1948; Black, 1953). It is therefore necessary to use a sampling frequency greater than $2F_m$ to minimise aliasing.

The voltage of the d.c. component in the spectrum of an unmodulated train of rectangular pulses is equal to the pulse height multiplied by the duty ratio of the pulse train (i.e. the ratio of pulse duration to pulse repetition period). If modulation causes changes in either pulse height or duty ratio, there is a corresponding modulation of the d.c.

component. Equation 1.55 shows that the spectrum of a PAM wave contains a voltage at the modulating frequency (f_m) whose ratio to the d.c. component ($\frac{1}{2}a_0$) is equal to the depth of modulation (m) of the pulse height. For PLM, the duty ratio varies with pulse duration, so the baseband frequency component is proportional to the depth of modulation of the pulse duration. In PFM, the duty ratio increases if more pulses are generated in a given period and decreases if fewer are generated. Consequently, a low pass filter can be used for demodulating PLM and PFM, just as for PAM.

In PPM, there is no change in pulse height or pulse duration and the mean number of pulses per unit of time is constant. Consequently, the spectrum has only a small component at the modulating frequency. PPM is therefore demodulated by converting it to PLM before low pass filtering.

In pulse time modulation, the pulse amplitude conveys no information. It is therefore possible to use a 'slicer' at the receiver, as shown in Figure 1.18, to produce pulses from which amplitude variations

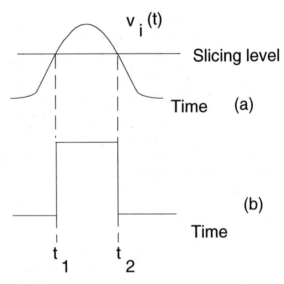

Figure 1.18　Use of slicer in pulse time modulation: (a) input pulse; (b) output pulse

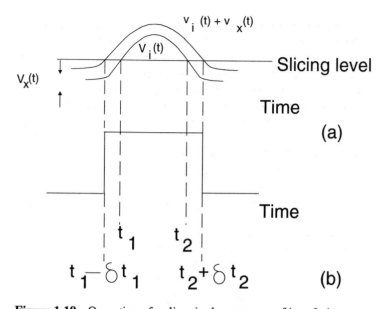

Figure 1.19   Operation of a slicer in the presence of interfering voltage (v_x): (a) input pulse; (b) output pulse

due to interference or noise have been removed. If the input pulses could have zero rise and fall times, no interference or noise would reach the output of the slicer. In practice, however, pulses have finite rise and fall times. As shown in Figure 1.19 interference or noise produces a displacement of the time at which the input pulse crosses the slicing level and thus of the time at which the output pulse appears from the slicer. This results in noise or interference in the demodulated output signal. This time displacement is given by Equations 1.58 and 1.59.

$$\delta t_1 = \left(\frac{dt}{dv_i} \right)_{t_1} v_x(t_1) \tag{1.58}$$

$$\delta t_2 = -\left(\frac{dt}{dv_i} \right)_{t_2} v_x(t_2) \tag{1.59}$$

These equations show that the demodulated output voltage varies inversely with the slope (dv_i/dt) of the input pulse. This slope is proportional to the bandwidth of the transmission path. Thus, the improvement in output signal to noise power ratio is proportional to the square of the bandwidth (Jelonek, 1947; Kretzmer, 1950; Black, 1953). It is assumed above that the noise voltage itself does not cross the slicing level; if it does, misoperation of the slicing circuit occurs and there is a severe deterioration of the output signal to noise ratio. Thus, there is a threshold phenomenon similar to that shown in Figure 1.15.

1.4.4 Time division multiplexing

Pulse modulation is used for time division multiplexing (TDM). It enables channels to share a wideband transmission path by using it at different times. The principle is shown in Figure 1.20. At the sending terminal, a baseband channel is connected to the common transmission path by means of a sampling gate which is opened for short intervals by means of a train of pulses. In this way, samples of the

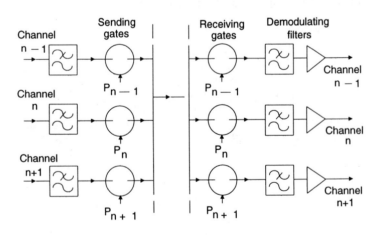

Figure 1.20 Elementary TDM system with one direction of transmission only

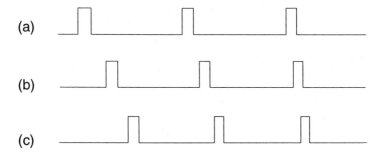

Figure 1.21 Channel pulser trains for the TDM system of Figure 1.20; (a) (n-1)th channel; (b) nth channel; (c) (n+1)th channel

baseband signal are sent at regular intervals by means of amplitude modulated pulses.

Pulses with the same repetition frequency f_r but staggered in time, as shown in Figure 1.21 are applied to the sending gates of the other channels. Thus the common transmission path receives interleaved trains of pulses, modulated by the signals of different channels. At the receiving terminal, gates are opened by pulses coincident with those received from the transmission path so that the demodulator of each channel is connected to the transmission path for its allotted interval and disconnected throughout the remainder of the pulse repetition period. The combination of a multiplexer and a demultiplexer at a TDM terminal is sometimes referred to as a muldex.

The pulse generator at the receiving terminal must be synchronised with that at the sending terminal to ensure that each incoming pulse train is gated to the correct outgoing baseband channel. A distinctive synchronising pulse signal is therefore transmitted in every repetition period in addition to the channel pulses. The complete waveform transmitted during each repetition period contains a number of time slots: one is allocated to the synchronising signal and the others to the channel samples. The complete waveform is called a frame, by analogy with a television signal waveform, and the synchronising signal is called the frame alignment signal.

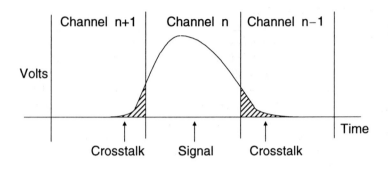

Figure 1.22 Inter-channel crosstalk in TDM systems

The elementary TDM system shown in Figure 1.20 uses PAM. Pulse length modulation and pulse position modulation can also be used. Pulse frequency modulation is unsuitable. If the PRF of one channel changes, its pulse will leave its allotted time slot and drift across the time slots of the other channels.

Attenuation and delay distortion of a transmission path cause pulse dispersion. The pulse of each channel spreads into the time slots of adjacent channels, as shown in Figure 1.22. This causes inter-channel crosstalk (Flood and Tillman, 1951; Flood, 1952). Consequently, analogue pulse modulation is now mainly used as an intermediate stage towards pulse code modulation (PCM), in order to obtain immunity to crosstalk by using digital transmission. Digital modulation methods are discussed in the next chapter.

1.5 Acknowledgement

This chapter contains material reproduced by permission of the Institution of Electrical Engineers from Flood, J.E. and Cochrane, P. (1991) *Transmission Systems*, Peter Peregrinus.

1.6 References

Black, H.S. (1953) *Modulation Theory*, Van Nostrand.

Brown, J. and Glazier, E.V.D. (1974) *Telecommunications* (2nd edition), Chapman and Hall.

Coates, R.F.W. (1975) *Modern Communication Systems*, Macmillan.

Fitch, E. (1947) *The Spectrum of Modulated Pulses*, J. Inst. Elect. Engrs. **94**, Pt. 111A, p. 556.

Flood, J.E. (1952), *Crosstalk to Time Division Multiplex Communication Systems Using Pulse Position and Pulse Length Modulation*, Proc. Inst. Elec. Engrs., **99**, Pt. IV, p. 64.

Flood, J.E. and Tillman, J.R. (1951) *Crosstalk in Amplitude Modulated Time Division Multiplex Systems*, Proc. Inst. Elec. Engrs., **98** Pt. III, p. 279.

Gardner, F.M. (1979) *Phaselock Techniques*, English Universities Press.

Gosling, W. (1986) *Radio Receivers*, Peter Peregrinus.

Hills, M.T. and Evans, B.G. (1973) *Transmission Systems*, Allen & Unwin.

Jahnke, E. and Emde, F. (1945) *Tables of Functions with Formulas and Curves*, Dover.

Jelonek, Z. (1947) *Noise Problems in Pulse Communication*, J. Inst. Elec. Engrs. **94**, Pt. IIIA, p. 533.

Kingdom, D.J. (1991) *Frequency Division Multiplexing*, Chap.5 in Flood, J.E. and Cochrane, P. (Eds) *Transmission Systems*, Peter Peregrinus.

Kretzmer, E.R. (1950) *Interference Characteristics of Pulse Time Modulation*, Proc. Inst. Radio Engineers, **38**, p. 252.

Langford-Smith, F. (1960) *Radio Designer's Handbook*, (4th edition), Iliffe.

Moss, S.H. (1948) *Frequency Analysis of Modulated Pulses*, Phil, Mag., **39**, p. 663.

Pappenfus, E.W., Bruene, W.B. and Schoenike, E.O. (1964) *Single Sideband Principles and Circuits*, McGraw-Hill.

Roberts, J.H. (1977) *Angle Modulation*, Peter Peregrinus.

Sturley, K.R. (1965) *Radio Receiver Design*, (3rd edition), Chapman & Hall.

Tucker, D.G. (1953) *Modulators and Frequency Changes for Amplitude Modulated Line and Radio Systems*, Macdonald.

Viterbi, A.J. (1966) *Principles of Coherent Communication*, McGraw-Hill.

2. Digital modulation

2.1 Introduction

The term digital modulation is used in two senses, both of which are discussed in this chapter:

1. Modulation of an analogue carrier by a digital baseband signal.
2. Modulation of a digital carrier by an analogue baseband signal.

A digital signal may modulate an analogue carrier by using amplitude modulation, frequency modulation or phase modulation, as discussed in Chapter 1. Examples are wireless telegraphy and the use of modems for transmitting data over analogue telephone circuits.

An important example of the use of a digital carrier for transmitting an analogue baseband signal is the use of pulse code modulation (PCM) for sending voice signals over digital circuits. This enables telephony to obtain the advantages of digital transmission, namely effective immunity to distortion, interference, crosstalk and noise and a constant transmission performance regardless of the length of a telephone connection and its routeing.

2.2 Digital transmission

In a digital transmission system, a modulator lies between the baseband input channel and the transmission path and a demodulator lies between that and the baseband output channel. The baseband input signal and the modulator determine the form of the transmitted signal and the demodulator is required to reconstruct the baseband signal with adequate fidelity. The overall channel between the modulator input and demodulator output must have the required properties and these will now be considered.

2.2.1 Bandwidth requirements

The minimum bandwidth needed to transmit a digital signal at B bauds has been shown by Nyquist (1928) to be $W_{min} = \frac{1}{2}B$ Hertz. This can be demonstrated as follows. Consider a binary signal consisting of alternate '0's and '1's. This produces a square wave of frequency $\frac{1}{2}B$. Let this be applied to an ideal low pass filter. If the cut off frequency of the filter is as in Equation 2.1 where ε is small, the output is a sine wave of frequency $\frac{1}{2}B$ and the original square wave can be recovered by sampling it at its positive and negative peaks.

$$\textit{Filter cut off frequency} = \frac{1}{2}B + \varepsilon \tag{2.1}$$

If the cut off frequency is reduced to the value given in Equation 2.2, the output consists only of the d.c. component and the signal is lost.

$$\textit{Cut off frequency} = \frac{1}{2}B - \varepsilon \tag{2.2}$$

Consequently, the minimum bandwidth required is given by Equation 2.3.

$$\frac{1}{2}B - \varepsilon < W_{min} < \frac{1}{2}B + \varepsilon \tag{2.3}$$

In the limit when ε tends to zero, Equation 2.4 is obtained.

$$W_{min} = \frac{1}{2}B \tag{2.4}$$

This result can also be demonstrated for the case when the transmitted symbols are very short pulses (which approximate to impulses), instead of the full width pulses considered above. The impulse response, h(t), of the ideal low pass filter of bandwidth W is given by Equation 2.5.

$$h(t) = \frac{\sin 2\pi W t}{2\pi W t} \tag{2.5}$$

This response has its maximum at t = 0 (where h(0) = 1) and is zero for t given by Equation 2.6 where T is as in Equation 2.7.

$$t = \pm n T \tag{2.6}$$

$$T = \frac{1}{2} W \tag{2.7}$$

If pulses are transmitted at rate B = 2W and each is detected by sampling at the time when it has its maximum output voltage, the outputs due to all preceding and following pulses are zero at that time, i.e. there is no intersymbol interference (ISI). Thus, it is possible to transmit pulses at rate B bauds through a channel of bandwidth W = ½B without any ISI.

In practice, it is not possible to obtain a channel with an ideal low pass characteristic. (Equation 2.5 shows that, if it were possible, an output voltage would appear before the input pulse is applied!) However, Nyquist showed that zero ISI can also be obtained if the gain of the channel changes from unity to zero over a band of frequencies with a gain frequency response that is skew symmetrical about f = ½B. It is also impossible to generate a perfect impulse (since it has zero duration and infinite amplitude). The transfer characteristic of the channel should therefore be equalised so that the output signal has the required spectrum. A commonly used signal is that having the raised-cosine spectrum given by Equation 2.8 to 2.10.

$$F(f) = 1 \quad \text{for } 0 \le f \le \frac{(1-\alpha)}{2T} \tag{2.8}$$

$$F(f) = \frac{1}{2} \left(1 + \sin \frac{\pi}{2\alpha} (1 - 2fT) \right)$$
$$\text{for } \frac{1-\alpha}{2T} \le f \le \frac{1+\alpha}{2T} \qquad (2.9)$$

$$F(f) = 0 \quad \text{for } f \ge \frac{1+a}{2T} \qquad (2.10)$$

This gain/frequency response rolls off sinusoidally from unity to zero in the frequency band from $\frac{(1-\alpha)}{2T}$ to $\frac{(1+\alpha)}{2T}$. Thus, for 100% roll off (i.e. $\alpha = 1$), the spectrum occupies a bandwidth of 1/T, which is twice the theoretical minimum requirement. Bandwidth is used most efficiently by using as small a roll off α as possible, but problems of timing and equalisation increase as α is reduced.

2.2.1.1 *Equalisation*

Digital transmission systems can use gain and phase equalisation to obtain an output signal spectrum corresponding to a pulse waveform with negligible intersymbol interference, e.g. the raised cosine spectrum described above. However, time domain equalisers are often employed.

A common form of time domain equaliser is the transversal equaliser (TVE) shown in Figure 2.1. This consists of a delay line tapped at intervals equal to the intersymbol interval T. Each tap is connected to an amplifier (which may be an inverter to obtain negative gain). The output of the equaliser is the sum of the outputs of these amplifiers. It is possible to adjust the gains of the amplifiers (in magnitude and sign) to cancel ISI by adding appropriately weighted versions of preceding and following pulses at the time of each symbol, and thus cancel interference between them.

A TVE can be adjusted manually. However, if the characteristics of the transmission path change with time, it is preferable for the equaliser to be adjusted automatically. A TVE which does this for itself during normal operation is called an adaptive equaliser (Lucky et al., 1968; Clark, 1985). Adaptive equalisers are used in data

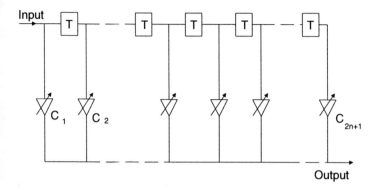

Figure 2.1   Block diagram of a transversal equaliser

modems which may be used on a variety of different connections in a switched telecommunication network. They are also used for digital radio links, which must cater for varying propagation conditions (de Belin, 1991).

So far, it has been assumed that the objective of equalisation is to eliminate intersymbol interference. An alternative is to have a large but well defined ISI. This is known as partial response equalisation (Lender, 1963; Kretzmer, 1966). It effectively turns a binary signal into a 3-level signal and enables a smaller bandwidth to be used. This can also be done digitally by combining adjacent pulses at the sending end of the transmission link. This form of partial response operation is called duobinary coding (Lucky et al., 1968; Taub and Schilling, 1986; Dorward, 1991).

2.2.2 Effects of noise

The principal advantage of PCM and other forms of digital transmission is that it is possible to obtain satisfactory transmission in the presence of very severe crosstalk and noise. In the case of binary transmission, it is only necessary to detect the presence or absence of each pulse. Provided that the interference level is not so high as to

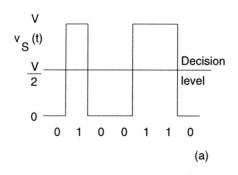

(a)

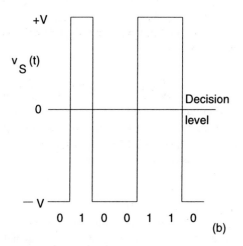

(b)

Figure 2.2 Detection of digital signals: (a) unipolar binary signal; (b) bipolar binary signal

cause frequent errors in making this decision, the output signal will be almost noise free.

Consider an idealised train of unipolar binary pulses, as shown in Figure 2.2(a). If the symbols '0' and '1' are equiprobable, i.e. as in Equation 2.11 the mean signal power corresponds to Equation 2.12. Thus, the signal to noise ratio is given by Equation 2.13, where σ is the r.m.s. noise voltage.

$$P(0) = P(1) = \frac{1}{2} \tag{2.11}$$

$$S = \frac{1}{2} V^2 \tag{2.12}$$

$$\frac{S}{N} = \frac{V^2}{2 \sigma^2} \tag{2.13}$$

The receiver compares the signal voltage v_s with a threshold voltage of $\frac{1}{2}V$, giving an output '0' when $v < \frac{1}{2}V$ and '1' when $v > \frac{1}{2}V$. If a noise voltage v_n is added, an error occurs if $v_n < -\frac{1}{2}V$ when $v_s = +V$, or if $v_n > +\frac{1}{2}V$ when $v_s = 0$. Thus, the probability of error, P_e, is given by Equation 2.14.

$$P_e = P(0) P(v_n > +\frac{1}{2} V) + P(1) P(v_n < -\frac{1}{2} V) \tag{9.14}$$

But Equation 2.15 holds and, for white noise, Equation 2.16 is obtained. Therefore Equation 2.17 is also obtained.

$$P(0) + P(1) = 1 \tag{2.15}$$

$$P(v_n < -\frac{1}{2} V) = P(v_n > +\frac{1}{2} V) \tag{2.16}$$

$$P_e = P(v_n > \frac{1}{2} V) \tag{2.17}$$

White noise has a normal or Gaussian probability density distribution, given by Equation 2.18 where σ is the standard deviation of v_n and is thus the r.m.s. noise voltage. Therefore Equation 2.19 is obtained.

$$p(v_n) = \frac{1}{\sigma \sqrt{2\pi}} \exp\left(\frac{-v_n^2}{2 \sigma^2} \right) \tag{2.18}$$

$$P(v_n > \frac{1}{2} V) = \int_{+\frac{V}{2}}^{\infty} p(v_n)\, dv_n = \frac{1}{2} - \int_0^{+\frac{V}{2}} p(v_n)\, dv_n$$

$$= \frac{1}{2} - \frac{1}{\sigma\sqrt{2\pi}} \int_0^{\frac{V}{2}} \exp\left(\frac{-v_n^2}{2\sigma^2}\right) dv_n \qquad (2.19)$$

Now the probability integral is defined as in Equation 2.20 and the complementary function is as in Equation 2.21. Therefore Equation 2.22 is obtained. Substituting from Equation 2.13 gives Equation 2.23.

$$Erf\, x = \frac{1}{\sigma\sqrt{2\pi}} \int_0^x \exp\left(\frac{-z^2}{2}\right) dz$$

$$(2.20)$$

$$Erfc\, x = \frac{1}{2} - Erf\, x \qquad (2.21)$$

$$P_e = P(v_n > \frac{V}{2}) = Erfc\, \frac{V}{2\sigma} \qquad (2.22)$$

$$P_e = Erfc\, \left(\frac{1}{2} \frac{S}{N}\right)^{\frac{1}{2}} \qquad (2.23)$$

Many authors use the form of error function defined as in Equation 2.24 and 2.25. Consequently Equations 2.26 and 2.27 are obtained.

$$erf\, x = \frac{2}{\sqrt{\pi}} \int_0^x \exp\left(-z^2\right) dz \qquad (2.24)$$

$$erfc\, x = 1 - erf\, x \qquad (2.25)$$

$$Erf\, x = \frac{1}{2} erf\left(\frac{x}{\sqrt{2}}\right) \qquad (2.26)$$

$$P_e = \frac{1}{2} \, erfc \left(\frac{1}{2} \sqrt{\frac{S}{N}} \right)$$

(2.27)

Tables of $erf \, x$ have been published (e.g. Jahnke and Emde, 1960; Betts, 1970). However, for $x > 3$, $erf \, x$ is so close to unity that it is usually not tabulated. In these cases, the asymptotic expression for the error function (Beckmann, 1967) should be used as in Equation 2.28.

$$erf \, x \approx 1 - \frac{\exp(-x^2)}{x \sqrt{\pi}}$$

(2.28)

This expression enables calculations to be made of the very low error probabilities obtained with good signal to noise ratios. The variation of error probability over a range of signal to noise ratios is shown in Figure 2.3.

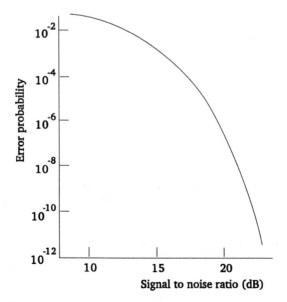

Figure 2.3   Error rate for transmission of unipolar binary signal disturbed by white noise

If the signal to noise ratio is calculated for the peak power (v^2/σ^2) instead of the mean power ($v^2/2\sigma^2$), then Equation 2.29 is obtained.

$$P_e = Erfc\left(\frac{1}{2}\sqrt{\frac{S}{N}}\right) = \frac{1}{2} erfc\left(\frac{1}{2}\sqrt{\frac{1}{2}\frac{S}{N}}\right) \tag{2.29}$$

If a bipolar binary signal is used, as shown in Figure 2.2(b), the mean signal power is given by Equation 2.30 and the signal noise ratio by Equation 2.31.

$$S = V^2 \tag{2.30}$$

$$\frac{S}{N} = \frac{V^2}{\sigma^2} \tag{2.31}$$

The receiver gives an output '0' when $v < 0$ and '1' when $v > 0$. An error will occur if $v_n < -V$ when $v_s = +V$, or if $v_n > +V$ when $v_s = -V$. Thus, the probability of error is given by Equation 2.32.

$$P_e = P(0)\,P(v_n > +V) + P(1)\,P(v_n < -V)$$

$$= P(v_n > +V)$$

$$= \int_V^\infty p(v_n)\,dv_n = \frac{1}{\sigma\sqrt{2\pi}}\int_V^\infty \exp\left(-\frac{v_n^2}{2\sigma^2}\right)dv_n$$

$$= Erfc\,\frac{V}{\sigma} = Erfc\,\sqrt{\frac{S}{N}} = \frac{1}{2} erfc\,\sqrt{\frac{1}{2}\frac{S}{N}} \tag{2.32}$$

Consequently, the same error rate is obtained with a 3dB lower signal to noise ratio. Alternatively, a much lower error rate can be obtained for the same signal to noise ratio.

The above analysis can be extended to multilevel digital signals, by replacing the pulse amplitude with the spacing between adjacent signal levels. If the number of levels is large, the error rate is nearly doubled. This is because all the intermediate levels can be misinterpreted in either direction, owing to noise voltages of either polarity.

For the case of a unipolar binary signal disturbed by white noise, the bit error rate, calculated from Equation 2.23, varies with the signal to noise ratio as shown in Figure 2.3. For example, if the signal noise ratio is 20dB, less than one digit per million is received in error. For telephone transmission, an error rate of 1 in 10^3 is intolerable, but an error rate of 1 in 10^5 is acceptable. Lower error rates are required for data transmission; if the error rate of the transmission link is inadequate, it is necessary to use an error detecting or error correcting code for the data (Lucky et al, 1968; Taub and Schilling, 1986).

On a long transmission link, it is possible to use regenerative repeaters instead of analogue amplifiers. A regenerative repeater (Dorward, 1991) samples the received waveform at intervals corresponding to the digit rate. If the received voltage at the sampling instant exceeds a threshold voltage, this triggers a pulse generator which transmits a pulse to the next section of the link. If the received voltage is below the threshold, no pulse is generated. If both positive and negative pulses are transmitted, the regenerator is required to determine whether the received voltage is positive or negative and to retransmit pulses of either polarity.

2.2.3 Optimal detection

In a digital transmission system, the receiver attempts, at the required instants of time, to determine whether a pulse is present that represents valid information or just noise. Optimum strategies for doing this have been developed.

If a digital signal uses symbols $s_1(t)$, $s_2(t)$,....., $s_r(t)$,.....,$s_n(t)$, each of duration T, these should ideally be orthogonal; i.e. there should be no correlation between them. This can be expressed mathematically as in Equation 2.33, where φ_{qr} is a cross correlation function and φ_{rr} is an autocorrelation function.

$$\varphi_{qr} = \int_0^T s_q(t)\, s_r(t)\, dt = 0 \quad \text{for } q \neq r$$
$$\neq 0 \quad \text{for } q = r \tag{2.33}$$

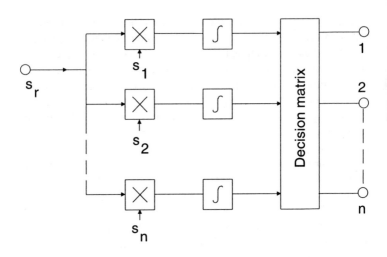

Figure 2.4   Correlation receiver

This suggests that an ideal receiver for the signal would be a bank of correlators, as shown in Figure 2.4. Each consists of a multiplier followed by an integrator. The signal is applied to one input of each multiplier and the waveform corresponding to one of the symbols is applied to the other. In the absence of noise, only one correlator will produce an output. If noise is present, other correlators will also produce some output. However, that corresponding to the symbol actually received should produce the largest output and this is selected as being the one most likely to be present. This maximum likelihood strategy has been shown to be the optimum for additive Gaussian noise when the symbols are equally probable (Harman, 1963).

Since an integrator is a low pass filter, each of the correlators in Figure 2.4 is equivalent to a coherent demodulator. Thus, a coherent demodulator with the appropriate waveform applied as its local carrier acts as a correlation detector.

It may be noted that the incoming pulses to a correlation receiver do not need to be a baseband signal. They can be pulses of a modulated carrier.

Another approach is to postulate that there can be a filter which enhances the signal and reduces the noise voltage as much as possible. Such a filter is called a matched filter, because its transfer function is tailored to the particular input signal to be detected.

It has been shown (Betts, 1970; Brown and Glazier, 1974; Schwartz, 1970; Taub and Schilling, 1986) that the impulse response of the required filter is given by Equation 2.34, where T is the duration of the signal.

$$h(t) = s_r(T - t) \tag{2.34}$$

The impulse response of the matched filter is thus the signal waveform with time reversed. It must be zero for $t > T$, at which time a decision is made as to whether $s_r(t)$ was present.

The output voltage, $v_0(t)$, due to the signal $s_r(t)$ is given by its convolution with $h(t)$ as in Equation 2.35.

$$v_o(t) = \int_0^t s_r(\tau) s_r(t - T - \tau) d\tau \tag{2.35}$$

Since this integral is zero outside the interval for which $s_r(t)$ and $s_r(t - T - \tau)$ exist, Equation 2.35 corresponds to the autocorrelation function φ_{rr} of $s_r(t)$. Consequently, the matched filter is equivalent to a correlation detector.

Since the impulse response of the matched filter is the time inverse of the input signal, if the latter is symmetrical in time, both are of the same shape. For example, if the signal is a pulse of (sin x)/x shape, so is the impulse response of the filter. The (sin x)/x pulse has a uniform spectrum up to frequency W, so the matched filter is an ideal low pass filter with cut off frequency W. This is hardly surprising, since it passes all the energy of the pulse and stops all noise outside the band of the signal. In the case of a sinusoidal carrier which is amplitude modulated by a (sin x)/x pulse, the matched filter will be an ideal band pass filter of bandwidth 2W, centred on the carrier frequency.

In a digital communication system, the input to the receiver should be band limited by a well designed filter. Consequently, the improvement obtained by using a matched filter or correlation detector is likely to be only about 1dB (Brown and Glazier, 1974).

2.3 Digital modulation of a carrier

2.3.1 Amplitude shift keying

If a binary signal is used to modulate the amplitude of a carrier to the greatest possible depth, the carrier is switched on and off. This is known as amplitude shift keying (ASK) or on off keying (OOK). The method is widely used in wireless telegraphy. Another important example is its use in optical communication systems. At the sending end, a binary electrical signal switches on and off a laser or light emitting diode. At the receiving end, a photodiode is switched on and off to reproduce the binary electrical signal.

In ASK, the spectrum of the transmitted signal consists of a component at the carrier frequency, together with an upper sideband which is a translated version of the baseband spectrum and a lower sideband which is an inverted version. It was shown earlier that the minimum bandwidth required to transmit the baseband signal is $W = \frac{1}{2}B$ (where B is the digit rate in bauds). It therefore follows that the minimum bandwidth required to transmit the ASK signal is $2W = B$. If the baseband signal has a raised cosine spectrum with 100% roll off, the bandwidth of the ASK signal is increased to 2B.

If the carrier amplitude is V_c, the power when it is switched on corresponds to $\frac{1}{2}V_c^2$. If the on and off states are equiprobable, the mean power is equal to half this value. The transmitted signal is a fully modulated carrier of amplitude $\frac{1}{2}V_c^2$, so that the power at the carrier frequency, f_c, corresponds to $\frac{1}{8}V_c^2$. Therefore, half of the total transmitted power is in the carrier and half is in the sidebands.

ASK can be demodulated by an envelope demodulator, as shown in Figure 1.5 (Chapter 1). However, at poor input signal/noise ratios, a coherent demodulator gives a better output signal/noise ratio and so a smaller bit error rate. Comparing Figures 1.6(b) (Chapter 1) and Figure 2.4 shows that the coherent demodulator is an optimum detector for the signal $s(t) = V_c \cos \omega_c t$.

For binary coded data, a coherent demodulator gives an output of +V for '1' and zero for '0'. The error rate thus corresponds to that of a unipolar baseband system and is given by Equation 2.27, where the signal/noise ratio is that at the output of the demodulator.

It is shown in Chapter 1 that the output signal/noise ratio, S_o/N_o, of the coherent demodulator is twice the input signal/noise ratio, S_i/N_i, when the latter is calculated from the sideband power alone. However, in this case, the sideband power is only half the total transmitted power. Therefore, $S_o/N_o = S_i/N_i$ and the error probability is given by Equation 2.27, where S/N is the input signal/noise ratio of the coherent demodulator.

The output voltage from a coherent demodulator is proportional to $\cos \theta$, where θ is the phase angle between the incoming carrier and the local carrier. Thus, ideal demodulation depends on keeping the local carrier in phase with the incoming carrier at all times.

This is not obtained when the carrier at the transmitter is switched on at arbitrary instants. Consequently, the carrier must be synchronised with the baseband signal and be an exact multiple of its digit rate.

If this is not possible, an envelope demodulator should be used to remove the effect of carrier phase variation. Use of a matched filter only improves the signal/noise ratio by about 1dB, so the difference between the input signal/noise ratios required for coherent and non-coherent ASK is only about 1dB (Bylanski and Ingram, 1987). It can be shown (Peebles, 1976) that the bit error probability for a binary non-coherent ASK system is approximately given by Equation 2.36.

$$P_e = \frac{1}{2} \left(1 + \frac{1}{\sqrt{\pi (S/N)}} \right) \exp \left(-\frac{1}{4} \frac{S}{N} \right) \qquad (2.36)$$

Single sideband suppressed carrier (SSBSC) transmission is unsuitable for digital data, because the baseband signal has a substantial low frequency content. Vestigial sideband (VSB) transmission (see Chapter 1) can be used. It enables a significant bandwidth saving to be obtained at high data rates (Bennett and Davey, 1965). For example, 9600 baud transmission can be obtained over telephone circuits with 8-level ASK using VSB transmission.

2.3.2 Frequency shift keying

If a digital signal is used to modulate a carrier in frequency, this is known as frequency shift keying (FSK). It is used in voice frequency telegraph systems (Chittleburgh et al., 1957) and for wireless telegraphy in the h.f. band (Hills and Evans, 1973). It is also widely used for data transmission over telephone connections in public switched telecommunication networks (Bennett and Davey, 1965; Lucky et al., 1968). A typical data modem for use on telephone circuits uses frequencies of 1.3kHz and 1.7kHz to transmit at 600 bauds or 1.3kHz and 2.1kHz to transmit at 1200 bauds (ITU-T, formerly CCITT, Recommendation V.23).

As shown in Figure 2.5, binary FSK is equivalent to applying ASK to two carriers of frequencies f_1 and f_2, one being switched on when

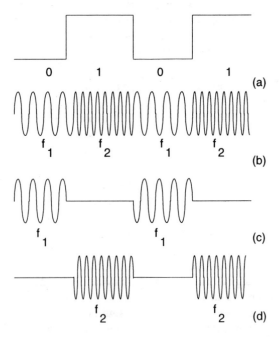

Figure 2.5 Frequency shift keying; (a) baseband signal; (b) FSK signal; (c) and (d) constituent ASK signals

the baseband signal is '0' and the other when it is '1'. In fact, some systems use two separate oscillators, instead of one whose frequency is shifted. However, this is not true frequency modulation, since there is no continuity of phase at the instants of frequency transition. It may be termed non-coherent FSK.

For each of the constituent ASK signals in Figures 2.5(c) and 2.5(d), half of the power lies in the sidebands and half in the carrier. Therefore the complete FSK signal has half of its power in sidebands and one quarter at f_1 and at f_2.

If the frequency deviation is large compared with the digit rate (high index f.m.), the resulting spectrum is that corresponding to amplitude shift keying of f_1 and f_2, with the energy concentrated about these two frequencies as shown in Figure 2.6(a). If the frequency deviation is reduced to become comparable to the digit rate (low index f.m.), the sidebands about f_1 and f_2 merge, as shown in Figure 2.6(b). Consequently, little more bandwidth is required than for ASK.

A high value of modulation index gives an improvement in signal to noise ratio, as in analogue systems. However, this may be unnecessary for a digital signal. Moreover, it would reduce the data

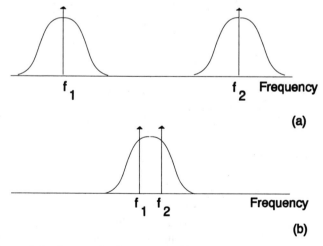

Figure 2.6 Spectra for frequency shift keying; (a) large freaquency deviation; (b) small frequency deviation

throughput obtainable in a given bandwidth. Sunde (1959) obtained a solution to satisfying Nyquist's minimum bandwidth condition for the case of FSK. He showed that, with certain assumptions, the minimum bandwidth is obtained when the frequency shift $(f_2 - f_1)$ is equal to the data rate in bauds. It is therefore common practice to use a frequency shift equal to the digit rate.

If the bandwidth occupied by the sidebands about f_1 and f_2 is approximately B and $f_2 - f_1 = B$, then the spectrum in Figure 2.6(b) occupies a bandwidth of approximately 2B. This is twice that needed for ASK. However, if coherent demodulation is used, it is possible to use a frequency shift of only 0.7B (Clark, 1983). In a multi-level FSK system, a similar separation is needed between the frequencies representing each of the levels. Thus, for a given symbol rate, the bandwidth increases with the number of levels.

An FSK signal can be demodulated by a frequency discriminator, as described in Chapter 1. Alternatively, a pair of filters tuned to f_1 and f_2 may be used with a pair of envelope demodulators. This is equivalent to two non-coherent ASK systems, so a similar error rate performance may be expected. It has been shown (Schwartz et al., 1966) that the error probability is approximately given by Equation 2.37.

$$P_e = \frac{1}{2} \exp\left(-\frac{1}{4} \frac{S}{N} \right) \tag{2.37}$$

If the phase of the FSK signal is coherent, a lower error rate can be obtained at low input signal/noise ratios by using two coherent demodulators, with local carriers f_1 and f_2 respectively. Since this is equivalent to a pair of coherent ASK systems, the error probability is the same and is given by Equations 2.23 and 2.27.

2.3.3 Phase shift keying

If phase modulation is used to transmit a digital signal, this is known as phase shift keying (PSK). If the signal is binary, the phase of the carrier is shifted between two positions which are 180° apart. This is sometimes known as phase reversal keying (PRK).

If a binary baseband signal, v_m, with values of -1 and $+1$ is applied to a product modulator to produce double sideband suppressed carrier (DSBSC) modulation of a carrier v_c given by Equation 2.38, the output voltage is given by Equation 2.39.

$$v_c = V_C \cos \omega_c t \tag{2.38}$$

$$\begin{aligned} v_o &= V_c \cos \omega_c t \quad \text{for } v_m = +1 \\ &= -V_c \cos \omega_c t = V_c \cos(\omega_c t + \pi) \quad \text{for } v_m = -1 \end{aligned} \tag{2.39}$$

Thus, binary PSK with a phase shift of 180° is identical with binary DSBSC transmission. At the receiver, a coherent demodulator with a local carrier $\cos \omega_c t$ will produce an output signal of $+1$ and -1.

Since PSK uses DSBSC amplitude modulation, it requires the same bandwidth as ASK and this is less than that needed for FSK. Moreover, since PSK is a suppressed carrier system, all the transmitted power is in the sidebands which convey the information. Compared with FSK and ASK which have only half of their power in the sidebands, this gives a 3dB advantage. Hence, the bit error probability is given by Equation 2.32. Consequently, for the same transmitted power, PSK gives a much lower error rate than FSK or ASK.

PSK signals may have more than two levels. For example, a 4-level PSK signal uses four carrier phases, separated by 90°. It is therefore sometimes known as quadrature phase shift keying (QPSK). In order to transmit a binary baseband signal, a pair of binary digits is combined to form a 4-level signal (00, 01, 10, 11) corresponding to the four phases transmitted. At the receiver, each of the four output values is used to generate two consecutive binary output digits. By this means, the digit rate on the transmission path and the required bandwidth are halved.

Alternatively, a higher digit rate can be achieved with a given bandwidth. Multiphase PSK is therefore used in digital radio systems for terrestrial links (de Belin, 1991) and for satellite communication systems (Evans, 1987; Nouri, 1991).

The penalty for the increased data rate obtained with multiphase PSK is a lower immunity to noise. Figure 2.7(a) shows that, for two

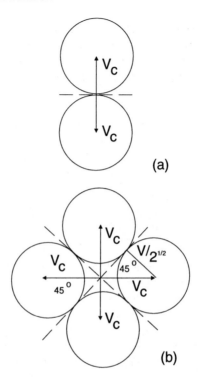

Figure 2.7   Signal-space diagrams for pulse phase modulation: (a) binary PSK; (b) 4-phase PSK

phase PSK, an interfering voltage must exceed V_c before an error can occur by detecting a '1' as '0' or vice versa. Figure 2.7(b) shows that, in four phase PSK, the tolerable interfering voltage is reduced by a factor of $1/\sqrt{2}$, i.e. 3dB. This is a disadvantage in the case of impulsive noise, such as that encountered on some switched telephone connections (Alexander et al., 1960; Williams, 1966). However, the bandwidth required for four phase PSK is only half that for two phase PSK. Consequently, in the case of white noise, the noise power is reduced by 3dB and the error rates are the same.

PSK requires a synchronous demodulator and synchronism is often difficult to maintain, particularly at high carrier frequencies. It is especially difficult to maintain if the transmission takes place through a fading medium or if either the transmitter or receiver is in a vehicle whose movement causes a Doppler shift of frequency. The difficulty can be overcome by transmitting a pilot carrier or by using a phase locked loop to control generation of the local carrier at the receiver (Viterbi, 1966).

The signal output voltage from a coherent demodulator is proportional to $\cos \theta$, where θ is the angle between the incoming and local carriers (see Chapter 1). Any departure from $\theta = 0$ reduces the amplitude of the output signal from the demodulator. However, there is no reduction in the noise output, so there is an increase in the error rate. Even if phase synchronism is correctly established initially, it may occasionally be lost. There can be a shift of 180° and this results in a sign inversion of all the output data.

The need for accurate carrier recovery can be avoided by using differential phase shift keying (DPSK). This does not require the generation of a local carrier at the demodulator. In DPSK, the information is conveyed by changes in phase between digits, instead of by the phase deviation of each from a reference carrier.

The principle of differential encoding is shown in Figure 2.8. The coder is fed with a sequence of binary symbols $s_1(1)$, $s_1(2)$, ..., $s_1(n)$, ..., where $s_1(n) = \pm 1$. Its output signal, $s_2(n)$, is obtained by multiplying $s_1(n)$ by the previous output from the coder, i.e. $s_2(n-1)$, resulting in Equation 2.40.

$$s_2(n) = s_1(n)\, s_2(n-1) \tag{2.40}$$

Thus, s_2 only changes sign if $s_1(n) = -1$.

Since $s_1(n)$ and $s_2(n-1)$ are single digit binary numbers, their multiplication can be carried out simply by means of an exclusive-OR logic gate.

Assuming error free transmission, the output signal from the decoder is given by Equation 2.41, regardless of whether $s_2(n-1)$ is $+1$ or -1 and whether the initial condition of s_2 was $s_2(0) = +1$ or $s_2(0) = -1$.

$$s_3(n) = s_2(n)\, s_2(n-1)$$

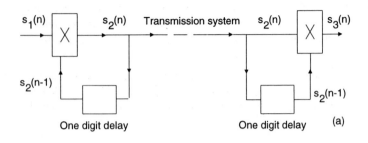

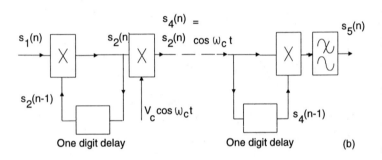

Figure 2.8   Principle of differential phase shift keying (DPSK); (a) differential coding and decoding; (b) DPSK system

$$= s_1(n)\, s_2(n-1)\, s_2(n-1) = s_1(n) \qquad (2.41)$$

In differentially encoded pulse phase modulation (DEPSK), the coder shown in Figure 2.8(a) is followed by a PSK modulator and the demodulator is followed by the decoder. However, the decoder can be incorporated in the coherent demodulator of the receiver, as shown in Figure 2.8(b). This is known as differential phase shift keying (DPSK). Then, the output signal of the demodulator is $s_5(n) = +1$ if $s_4(n)$ and $s_4(n-1)$ are represented by carriers of the same phase and $s_5(n) = -1$ if the phase has shifted by 180° between $s_4(n-1)$ and $s_4(n)$. However, the phase of s_4 will only have shifted if s_2 has changed in polarity as a result of $s_1(n) = -1$. Therefore Equation 2.42 is obtained.

$$s_5(n) = -1 \quad when \quad s_1(n) = -1$$

$$s_5(n) = +1 \quad when \quad s_1(n) = +1 \tag{2.42}$$

Although DPSK has the advantage of no errors due to loss of synchronism, the probability of error due to noise is greater than for PSK. Each input digit to the demodulator contributes to determining the values of two output digits. Consequently, a noise peak that would produce a single error with PSK can produce two errors with DPSK.

Calculation of the bit error probability for DPSK is complicated. It has been shown (Schwartz et al., 1966) that the error probability is given by Equation 2.43.

$$P_e = \frac{1}{2} \exp\left(-\frac{1}{2} \frac{S}{N} \right) \tag{2.43}$$

2.3.4 Comparative performance of ASK, FSK and PSK

Figure 2.9 shows, for signals perturbed by added Gaussian noise, how the bit error probability varies with the signal to noise ratio for binary systems using coherent and non-coherent ASK, coherent and non coherent FSK, PSK and DPSK. These curves were obtained from Equations 2.27, 2.32, 2.36, 2.37 and 2.43. For comparison, the error probabilities for unipolar and bipolar baseband transmission (from Equations 2.27 and 2.32) are also shown.

None of the modulation schemes, except PSK, gives as good an error performance as bipolar baseband transmission. Coherent ASK and FSK both achieve as good a performance as unipolar baseband transmission. Non coherent ASK and FSK are about 1dB worse than the corresponding coherent systems. DPSK is about 1dB worse than PSK.

Determination of the error probabilities for multilevel systems is more complex and will not be considered here. The reader is referred to Arthurs and Dym (1962), Bylanski and Ingram (1980) and Taub and Schilling (1986). For 4-level systems, it is found that coherent

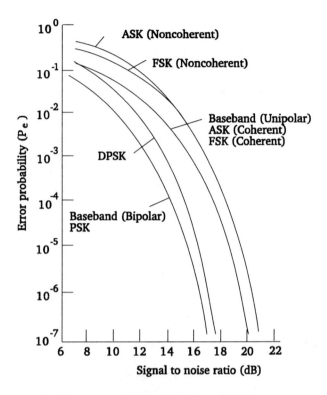

Figure 2.9 Bit error probability for various digital systems

FSK nearly equals the performance of PSK and ASK gives the worst performance (Bylanski and Ingram, 1980). For more than four levels, a better performance is obtained by using the hybrid modulation schemes described in the next section.

2.3.5 Quadrature amplitude modulation

If a carrier is modulated by a digital signal, its amplitude and phase can only have a finite number of values. These can be represented in a signal space diagram, as shown in Figures 2.10 and 2.11. These show an 8-level ASK signal and an 8-level PSK signal. The circles

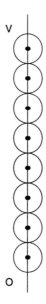

Figure 2.10   Signal-space diagram for eight state amplitude modulation

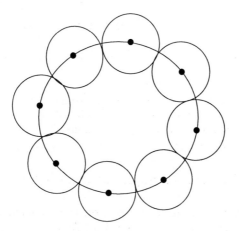

Figure 2.11   Signal-space diagram for eight state phase modulation

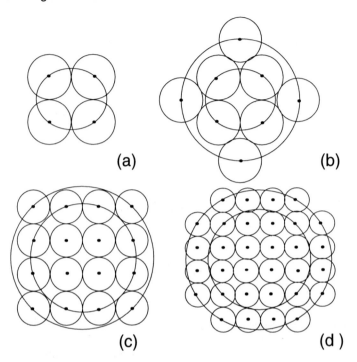

Figure 2.12   Signal-space diagrams for m-state quadrature amplitude modulation; (a) four state; (b) eight state; (c) sixteen state; (d) thrity-two state

show the maximum permissible perturbation from the ideal signal before an error may occur in the transmission. It is obvious that, in this respect, PSK is superior to ASK.

It is possible to combine ASK and PSK to obtain hybrid modulation, as shown in Figure 2.12. The 8-level scheme shown in Figure 2.12(b) has a similar error threshold voltage to the PSK scheme of Figure 2.11. However, it has a lower mean transmitted power, because four of the eight states have a smaller carrier amplitude than those in Figure 2.11.

These hybrid modulation schemes can be implemented using quadrature amplitude modulation (QAM), i.e. modulation of two

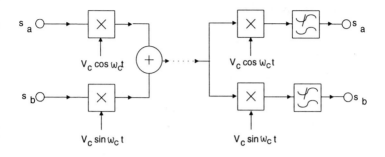

Figure 2.13 Quadrature amplitude modulation (QAM) system

carriers in quadrature as shown in Figure 2.13. If a 2-level signal is applied to each modulator, the output has a constant amplitude, as shown in Figure 2.12(a). It therefore produces quadrature phase shift keying (QPSK).

If a 3-level signal ($s_1 = -1$, $s_2 = 0$, $s_3 = +1$) is applied to each of the two modulators, the sum of their outputs has three values of its in phase component and three values of its quadrature component. This permits duobinary partial response coding and the system is known as a quadrature partial response (QPRS) system (Taub and Schilling, 1986).

If a 4-level signal is applied to each of the two modulators, the sum of their outputs has four values of its in phase component and four values of its quadrature component. This produces the sixteen state signal constellation shown in Figure 2.12(c). The demodulators reproduce the two 4-level signals at their outputs.

Four state QAM is used to produce QPSK to transmit data at 2400bit/s over switched telephone circuits with a signalling rate of only 1200 bauds (ITU-T Recommendations V.26 and V.27). Sixteen state QAM is used to provide 2400bit/s data transmission over switched telephone circuits with a signalling rate of only 600 bauds (ITU-T Recommendation V.22 bis) and 9.6kbit/s over private circuits with a signalling rate of only 2400 bauds (ITU-T Recommendation V.32).

2.4 Pulse code modulation

Pulse code modulation (PCM) can be produced by applying a train of amplitude modulated pulses to an analogue to digital (A/D) converter, as shown in Figure 2.14. Each analogue sample is thus converted to a group of on/off pulses which represents its voltage in a binary code. At the receiving terminal, a digital to analogue (D/A) converter performs the decoding process. The combination of a coder and decoder at a PCM terminal is often referred to as a codec. The group of bits (i.e. binary pulses) representing one sample is called a word or a byte. An 8-bit byte is sometimes called an octet. In a time division multiplex (TDM) system, the coder and decoder are required to perform their operations within the time slot of one channel. They can therefore be common to all the channels of a TDM system, as shown in Figure 2.14.

For telephony, speech samples are usually encoded in an 8-bit code. Since sampling is at 8kHz, a telephone channel requires binary digits to be sent at the rate of $8 \times 8 = 64$kbit/s. It was shown earlier that the minimum bandwidth required to transmit pulses is half the pulse rate. Thus, a bandwidth of at least 32kHz is required to transmit a single telephone channel. The advantages of digital transmission are won at the expense of a much greater bandwidth requirement.

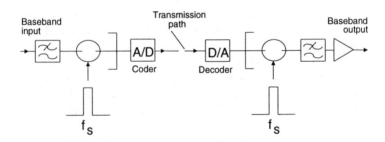

Figure 2.14   Block diagram of a PCM system (one direction of transmission only)

2.4.1 Sampling

In order to produce a digital signal, an analogue signal must first be sampled by a periodic pulse train to convert it from continuous to discrete time. If a continuous signal has a bandwidth W, then the minimum sampling frequency, f_s, that can be used is $f_s = 2W$. This is called the Nyquist rate. Subsequent removal of all frequency components except the original baseband by a low pass filter at the receiver recovers the original signal.

If the sampling rate is less than 2W, the lower sideband about f_s overlaps the baseband and it is impossible to separate them. This is called aliasing. It can be avoided by passing the input signal through a low pass filter before sampling it. It is essential to include such an anti-aliasing filter at the input of a sampling system. Since the nominal band of telephone signals is from 300Hz to 3.4kHz and the internationally agreed sampling frequency is 8kHz, there is a guard band of 1.2kHz between the top of the baseband and the bottom of the lower sideband of the sampling frequency. This must accommodate the transition between the pass band and the stop band of the anti-aliasing filter at the sending end and of the demodulating filter at the receiving end.

Anti-aliasing filters can be implemented in many ways. High order passive LC filters have often been used, offering low distortion and low noise, but with the physical drawbacks of large size and weight, and inevitable insertion loss. A more up to date approach might be the use of a Sallen-and-Key active filter (1955), which is smaller, lighter and loss free, but adds more noise, and can be more difficult to trim than the corresponding LC filter. Switched capacitor filters (Sedra and Smith, 1982) can be highly integrated, requiring only off chip capacitors, and hence lead to a cheap package which does not require large amounts of PCB area. They do, however, add considerably more noise than the passive and active filters.

The analogue sampler, commonly known as the 'sample-and-hold', performs the task of sampling the input analogue voltage and maintaining that voltage until the next sampling instant. A simple sample-and-hold is illustrated in Figure 2.15. When the Sample command is given, the buffer is connected via the FET switch to the

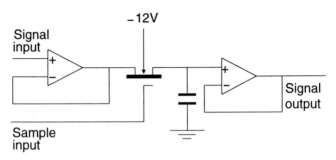

Figure 2.15   Simple sample-and-hold circuit

storage capacitor, the voltage on which tracks the input voltage thereafter. The capacitor is at all times buffered by a voltage follower. When the Sample command is removed, the capacitor is isolated from the input buffer. The voltage across the capacitor, and hence the voltage follower, will then remain steady until the next Sample command.

Such a simple design is marred by a number of adverse properties, so more complex designs are sometimes used (Wallace et al., 1991).

2.4.2 The coder

The analogue sampler performs the transformation from continuous time to discrete time representation. The coder or analogue to digital converter (ADC) performs the complementary operation of transformation from continuous amplitude to discrete amplitude. There are a variety of techniques which may be used in an ADC, such as dual slope conversion, successive approximation and flash conversion (Wallace et al., 1991).

The successive approximation ADC is built from a digital to analogue converter, a comparator, a storage register and controlling logic, as shown in Figure 2.16. This type of converter essentially operates by performing a binary search through the ADC dynamic range to find the input voltage.

Upon receiving the Convert command, the storage register is cleared. The output from the storage register is passed to the digital

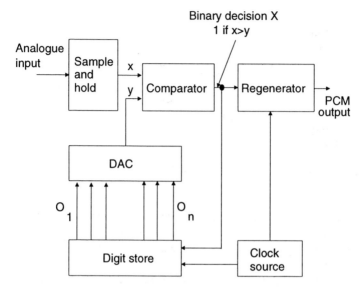

Figure 2.16 Successive approximation ADC

to analogue converter (DAC), but with the most significant bit set. This causes the DAC to output its half range voltage. This voltage is then compared with the input sample by the comparator. The comparator output is then stored in the most significant bit (MSB) of the storage register. The converter has thus decided whether the input voltage is in the lower or upper half of the ADC's dynamic range.

On the next clock cycle, the storage register output is passed to the DAC with its 2nd most significant bit set. The comparator output then indicates whether the input signal lies in the upper or lower half of that portion of the dynamic range selected in the previous decision. The comparator output is stored in the 2nd most significant bit of the register.

This process continues, passing through as many cycles as there are bits in the storage register, with each pass determining whether the input voltage lies in the upper or lower half of the range selected in the previous cycle. The storage register then contains the ADC conversion value.

The successive approximation ADC must perform several operations in sequence to encode each sample. For high sampling rates (e.g. above 1MHz), in order to reduce the conversion time, it is necessary to use a circuit which requires fewer operations to be performed.

The flash converter, shown in Figure 2.17, consists of a chain of fast comparators, with one input of each comparator driven by the

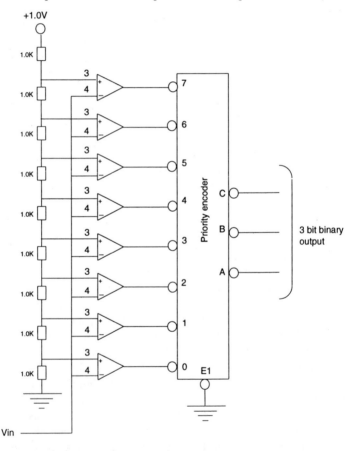

Figure 2.17   A 3 bit flash converter

voltage to be converted. The other input of each successive comparator is connected to the corresponding point of a resistor chain, containing as many resistors of equal value, as there are comparators.

The number of comparators, and hence resistors, in the divider chain, is equal to the number of levels which must be coded. For example, for the case of a 4-bit flash converter, capable of coding voltages in the range 0 to 1 V, 16 comparators are needed, connected to a resistor chain containing 16 equal resistors, connected across a one volt reference. The comparators thus compare the input voltage with $1/16$V, $1/8$V, $3/16$V, and so on.

The comparator outputs are then passed to a priority encoder, which determines the position of the last comparator in the chain to indicate that its reference voltage is lower than the flash encoder input voltage. This indicates directly which segment of the converter's dynamic range contains the input signal and gives the corresponding binary code output.

The main feature of the flash converter is its extremely rapid conversion time. They are often used for video ADC's. or any other application which requires a sample rate in excess of 1MHz.

The penalty for the speed is the rapid increase in complexity, and therefore power consumption and cost, as the desired number of bits increases. Indeed, it is rare to find a flash converter operating to more than 8-bit accuracy because of the cost of an array of more than 256 comparators.

Variation of input offset voltage between adjacent comparators can set a further limit to ultimate resolution. However, analogue and digital pipelining can be combined to implement two stage flash conversion, giving accuracy up to twice the number of bits with the same sample rate, and only twice the conversion time.

The above converters encode voltages between ground and some upper limit. It is more usual to want to convert voltages of both polarities.

This can be achieved either by:

1.   The addition of a voltage equal to half the upper limit, with two's complement coding of the signal then easily achieved by inversion of the resulting MSB.

2.   The use of a precision full wave detector, which results in an output signal of fixed polarity, together with information about whether the signal is positive or negative.

The latter scheme results naturally in the production of a 'sign-plus-magnitude' code.

2.4.3 The decoder

After the digitally encoded signal has been transmitted, it is often necessary to reconstruct an analogue signal at the receiver. Conversion from digital to analogue is generally easier to achieve than conversion from analogue to digital. There are again many techniques which may be used for this conversion, each possessing its own blend of merits and demerits.

The simplest form of DAC is shown in Figure 2.18. It consists of an array of resistors connected via an analogue switch either to a

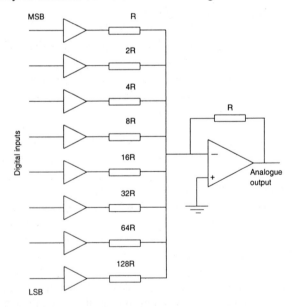

Figure 2.18 Simple digital to analogue converter

reference voltage, if the corresponding bit in the digital word is 'one', or to ground otherwise. The resistors are all connected to the virtual earth input of a current summing amplifier, and their values are chosen such that the currents flowing through the resistors when the corresponding bits are 'one' are binary weighted.

The output voltage of the summing amplifier is proportional to the sum of the input currents, and hence to the numerical value of the digital codeword.

The circuit illustrated in Figure 2.18 is simple, but it suffers some drawbacks. The linearity and monotonicity are both critically dependent on the precision of the resistors used. Any errors in resistor value will show particularly around the transition between the lower and upper halves of the converter's dynamic range. The resistor values will also almost certainly drift slowly with time and temperature. More elaborate designs based on the resistor ladder network overcome this drawback (Wallace et al., 1991).

2.4.4 Quantisation distortion

The digital signal can only represent a finite number of input signal voltages. For example, a binary code of k digits can only represent 2^k different input voltages. When the signal is reconstructed at the receiver, it is an approximation to the original signal, since it can only change in discrete voltage steps. This signal is said to be quantised, and a form of non-linear distortion, known as quantisation distortion, has been introduced.

The difference between the original signal and the reconstructed signal is the error voltage. Because the error voltage varies randomly with time, it is often referred to colloquially as 'quantisation noise'. Indeed, in PCM telephony, it sounds just like noise. If the acceptable r.m.s. noise voltage in an analogue signal is σ, then a discrete representation with a voltage level spacing of the order of σ will be adequate to represent the analogue signal.

Obviously, the closer the spacing of the discrete levels, the smaller will be the quantisation distortion. However, the smaller the spacing, the greater is the number of levels. This increases the number of digits in the code words representing the levels (e.g. doubling the number of levels increases the length of a binary code word by one bit). An increase in the number of digits representing every sample increases the bandwidth required to transmit them.

Consider a quantiser where the ith interval is centred around Q_i and possesses width Δ_i. Any input signal x in the range given by Equation 2.44, will be represented by the quantised amplitude Q_i.

$$Q_i - \frac{1}{2}\Delta_i \leq x < Q_i + \frac{1}{2}\Delta_i \tag{2.44}$$

The squared quantisation error will clearly be $(x - Q_i)^2$. Let us also assume that the probability density of the input signal is p(x). Therefore, to obtain the mean square error over the ith range, the squared error value must be averaged over all possible values of the input signal for that quantisation interval, as in Equation 2.45.

$$E_i^2 = \int_{Q_i - \frac{1}{2}\Delta_i}^{Q_i + \frac{1}{2}\Delta_i} p(x)(x - Q_i)^2 \, dx \tag{2.45}$$

If the signal varies widely in comparison to the size of a single quantisation interval, it can be assumed that p(x) is constant across that quantisation interval, having value $\frac{1}{\Delta_i}$. Therefore Equation 2.46 is obtained.

$$E_i^2 = \frac{1}{\Delta_i}\int_{Q_i - \frac{1}{2}\Delta_i}^{Q_i + \frac{1}{2}\Delta_i} (x - Q_i)^2 \, dx = \frac{1}{\Delta_i}\frac{\Delta_i^3}{12} = \frac{\Delta_i^2}{12} \tag{2.46}$$

We may now calculate the mean square noise over the entire quantiser dynamic range by averaging E_i over all values of i as in Equation 2.47, where P_i is probability of the signal voltage being in the ith interval and N is total number of quantisation intervals in the entire converter dynamic range.

$$E^2 = \frac{\sum\limits_{i=1}^{N} P_i E_i^2}{\sum\limits_{i=1}^{N} P_i} = \frac{1}{12} \sum_{i=1}^{N} P_i \Delta_i^2 \tag{2.47}$$

For the important case of a uniform quantiser, where all intervals have the same width (Δ) Equation 2.48 is obtained.

$$E^2 = \frac{\Delta^2}{12} \sum_{i=1}^{N} P_i = \frac{1}{12} \Delta^2 \tag{2.48}$$

For a sinsuoidal input signal, the maximum amplitude which can be handled is given by Equation 2.49, so the signal power is given by Equation 2.50.

$$V_m = \frac{1}{2} N \Delta \tag{2.49}$$

$$S = \frac{1}{2} V_m^2 = \frac{1}{8} N^2 \Delta^2 \quad volts \tag{2.50}$$

Therefore, the signal/quantisation noise ratio (SQNR) is as in Equation 2.51.

$$SQNR = \frac{S}{E^2} = \frac{3}{2} N^2 \tag{2.51}$$

Expressing this in decibels gives Equation 2.52.

$$SQNR = 1.8 + 20 \log_{10} N \quad dB \tag{2.52}$$

Now, for binary encoded PCM, $N = 2^k$ where k is the number of bits in the codeword. Substituting in Equation 2.51 gives Equation 2.53.

$$SQNR = 1.8 + 20\,k\log_{10} 2 \quad dB$$
$$= 1.8 + 6\,k \quad dB \tag{2.53}$$

Equation 2.53 shows that adding one digit to the code improves the SQNR by 6dB. Since the quantisation noise level is independent of signal level, a signal whose level is xdB lower than the maximum will have a SQNR which is xdB worse than that given by Equations 2.52 and 2.53.

2.4.5 Companding

In typical telephone speech, the probability distribution of signal amplitudes has a maximum at zero amplitude and becomes progressively smaller for high amplitudes, having a near Gaussian distribution. In addition, substantial level variations exist between one conversation and another, depending on such factors as the line length between the subscriber and the exchange, the sensitivity of the subscriber's handset and the loudness of the talker. These factors can lead to a variation of up to 30dB between one speaker and another (Purton, 1962). There is, therefore, a requirement for a wide dynamic range to be coded by the ADC for a speech signal.

A uniform quantiser has a constant noise power, independent of the signal level. Equation 2.53 shows that to code a signal with a dynamic range of 50dB, maintaining a signal to noise ratio of at least 30dB over that dynamic range, would require a 13-bit uniform quantiser. The ear, however, is a very non-linear sensor and, in the presence of a speech signal, any broadband noise which is 30dB quieter than the signal is masked by that speech. So, why use a uniform quantiser with its fixed noise power, unaffected by speech level? What is really needed is a quantiser which has a constant signal to noise ratio of 30dB over a 50dB dynamic range.

The signal/quantising noise ratio for small signals could be improved by compressing the level range before coding and using expansion after decoding at the receiver (i.e. companding).

It is not necessary to use a 'syllabic' compander, because the process can be performed independently on each speech sample. Such an 'instantaneous' compressor could be implemented by a

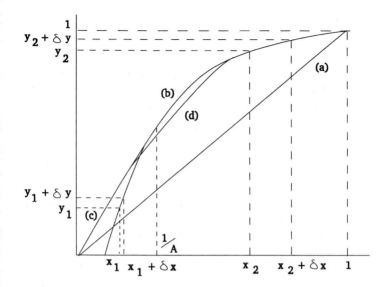

Figure 2.19 Compression characteristics: (a) linear law; (b) logarithmic law; (c) A-law; (d) μ-law

non-linear characteristic for which constant increments (δ_y) in output voltage correspond to input voltage increments (δ_x) which are proportional to the input voltage x, as shown in Figure 2.19(b). Thus Equations 2.54 and 2.55 are obtained.

However, log 0 = −∞ and a practical ADC must give zero output for zero input. A compression law is therefore needed which is logarithmic for large values of x but corresponds to a straight line through the origin for small values of x, as shown in Figures 2.19(c) and (d).

$$d y = \left(\frac{b}{x} \right) d x \qquad (2.54)$$

$$y = b \log cx \qquad (2.55)$$

One compression law that is used is the A-law shown in Figure 2.19(c). This is given by Equation 2.56.

$$y = \frac{1 + \log_e A x}{1 + \log_e A} \quad \text{for } \frac{1}{A} \leq x \leq 1$$

$$= \frac{A x}{1 + \log_e A} \quad \text{for } 0 \leq x \leq \frac{1}{A} \tag{2.56}$$

The section below $x = 1/A$ is linear. The section between $1/A$ and 1 is logarithmic and, for $A = 87.6$, this gives constant SQNR over a range of $20 \log_{10} 87.6 = 38$dB. For small signals, the A-law gives Equation 2.57.

$$\frac{dy}{dx} = \frac{A}{1 + \log_e A} \tag{2.57}$$

However, for uniform quantisation, $dy/dx = 1$. Thus, if the A-law companding advantage is defined as the improvement in SQNR compared with uniform quantisation, it is as in expression 2.58. For $A = 87.6$, this is 24dB.

$$20 \log_{10} \left(\frac{dy}{dx} \right) \tag{2.58}$$

Another compression law that is used is the μ-law shown in Figure 2.19(d). This is given by Equation 2.59.

$$y = \frac{\log_e (1 + \mu x)}{\log_e (1 + \mu)} \quad \text{for } 0 \leq x \leq 1 \tag{2.59}$$

This approximates to the logarithmic law for $x \geq \frac{1}{\mu}$ and to a linear law for $x << \frac{1}{\mu}$ since Equation 2.60 holds.

$$\log_e (1 + \mu x) = x - \frac{\mu^2 x^2}{2} + \frac{\mu^3 X^3}{3} - \ldots \tag{2.60}$$

From Equation 2.59, the slope of the characteristic is as in Equation 2.61.

$$\frac{dy}{dx} = \frac{1}{\log(1+\mu)} \frac{\mu}{1+\mu x} \tag{2.61}$$

Equation 2.61 shows that, when $\mu = 255$, the SQNR changes by less than 3dB over an input level change of 40dB. For small signals Equation 2.62 is obtained.

$$\frac{dy}{dx} = \frac{\mu}{\log_e(1+\mu)} \tag{2.62}$$

This corresponds to the companding advantage; thus, for $\mu = 255$, this is 33dB.

Early PCM systems use non-linear networks in conjunction with uniform encoding. Later systems used piecewise linear approximations to the compression law, since these are easy to implement digitally. Two such logarithmic quantising schemes for 8-bit encoding have been defined by ITU-T Recommendation G.711. They correspond respectively, to the A-law with A=86.7 and the μ-law with $\mu = 255$. The A-law is used across Europe, and the μ-law is used across America and Japan. Both schemes use codewords which bear a remarkable similarity to 8-bit binary floating point numbers. One bit is reserved for the sign, three bits for the exponent, and four bits for the mantissa, with the leading bit (which is always one) removed. The A-law characteristic is shown in Figure 2.20.

2.5 Differential pulse code modulation

2.5.1 Basic differential PCM

Conventional PCM permits any sample to have any value. The system can therefore cope with adjacent samples having values at the extreme opposite ends of the range. For example, it can transmit a signal of maximum amplitude at the maximum baseband frequency.

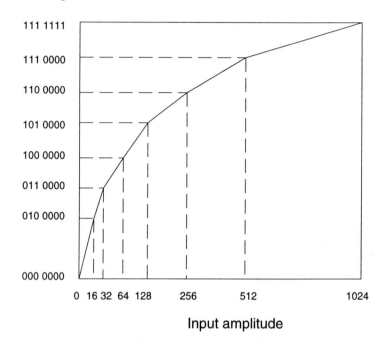

Figure 2.20 ITU-T (formerly CCITT) A-law characteristics

Such rapidly changing signals rarely occur, so it should be possible to convey the baseband signal with adequate fidelity by transmitting fewer bits per second than is needed by conventional PCM. Differential pulse code modulation (DPCM) does this by encoding the difference between each sample and the previous one, instead of encoding the actual value of each sample (Peebles, 1976; Taub and Schilling, 1986).

The block diagram of a basic DPCM system is shown in Figure 2.21. Each sample presented to the coder is the difference between the present value of the baseband input voltage and the integrated value of previous transmitted samples. At the receiver, the output of the decoder is fed to the baseband output channel via a similar integrator. Thus, the coder at the sending end is constantly updating the output of the receiver's integrator by comparing the output of its

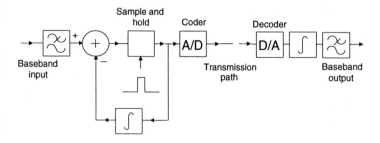

Figure 2.21 Block diagram of a differential PCM system

own integrator with the voltage received from the baseband input channel and transmitting the difference. Usually, the difference will be much smaller than the full amplitude of the input signal, so it can be encoded with fewer digits. Typically, 4-digit coding can be used instead of 8-digit coding, thereby halving the transmitted digit rate and the bandwidth required.

Occasionally, the input signal may change so rapidly that the difference between it and the integrator output exceeds the maximum voltage that can be encoded. This is known as slope overload. There is then an error in the output voltage from the receiver. However, this rapid rate of change is unlikely to persist, so the receiver output will be corrected over the next few samples.

2.5.2 Adaptive differential PCM

The basic DPCM system of Figure 2.21 may be thought of as predicting that each sample will not change from the previous one and transmitting the difference between each sample and this predicted value. Thus, the integrators at the sending and receiving ends are a simple form of predictor. The performance of the system can be improved, or the digit rate reduced, by using a more sophisticated form of predictor based on the a priori knowledge that the baseband signal to be encoded is a speech signal.

The predictor can be a self adaptive filter based on a model of the vocal tract (Wallace et al., 1991). This removes a large amount of the redundancy contained in voiced sounds, thereby reducing the magnitude of the difference signal to be encoded. This scheme is known as adaptive differential pulse code modulation (ADPCM). It enables a SQNR of about 40dB to be obtained with 4-bit encoding. Telephone transmission can therefore use only 32kbit/s instead of 64kbit/s.

The ITU-T has specified an algorithm for 32kbit/s ADPCM (Recommendation G.721). This is embodied in a transcoder, whose input is a normal 64kbit/s PCM signal (with A-law or μ-law companding) and whose output is 32kbit/s ADPCM. Block diagrams of this encoder and the corresponding decoder are shown in Figures 2.22 and 2.23.

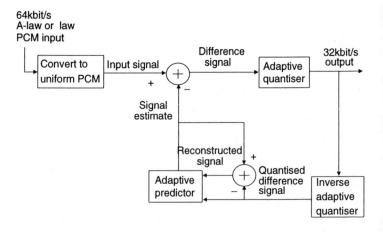

Figure 2.22 ADPCM encoder

Alternatively, ADPCM can be used to enable a 64kbit/s digital channel to transmit speech having a bandwidth of 7kHz instead of only 3.4kHz (ITU-T Recommendation G.722).

2.5.3 Linear predictive coding

Linear predictive coding (LPC) takes the idea of using a model of the vocal tract one stage further (Atal and Schroeder, 1970). A short block

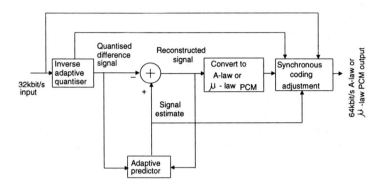

Figure 2.23 ADPCM decoder

of speech, typically 20ms in duration, is processed to extract pitch period and formant information, which describe respectively the activity of the vocal chords and the resonances of the vocal tract. Then, rather than transmitting speech samples, the LPC coder sends the pitch period and formant model coefficients for the block of speech. The decoder uses this information to reconstruct the speech block from the model.

Because of the small amount of information transmitted, LPC coders can use digit rates as low as 4.8kbit/s, although error protection of the code is essential.

However, LPC coders have the severe disadvantage of introducing a coding delay of two or three times the length of the speech block. Consequently, echo suppressors or echo cancellers must be used with LPC.

The Group Speciale Mobile (GSM), which was formed by the European PTTs, had the task of specifying the speech encoding algorithm for a pan European digital cellular radio system. It has chosen a variant of LPC, known as regular pulse excitation long term prediction (RPE-LTP) (ETSI, 1989). This will use a speech block of 20ms and will operate at 13kbit/s, with an additional 3kbit/s for error protection.

2.6 Delta modulation

2.6.1 Basic delta modulation

The limiting case of DPCM is when the number of bits in the code is reduced to only one. This is possible if over sampling is used, i.e. the sampling frequency is made much greater than the Nyquist rate. The difference that can occur between adjacent samples is then reduced, so that it rarely exceeds one quantum step. The method is called delta modulation (DM) (Steele, 1975).

A basic DM system is shown in Figure 2.24. At the sending end, the comparator causes the pulse generator to transmit +V when the input voltage exceeds the output voltage from the integrator and –V when it is less. As well as being transmitted, each of these pulses causes the output voltage from the integrator to rise or fall by one quantum step, so that it closely tracks the input voltage.

At the receiving end, each incoming positive or negative pulse causes the output voltage from a similar integrator to rise and fall by one quantum step, thereby reproducing the input signal of the sending end.

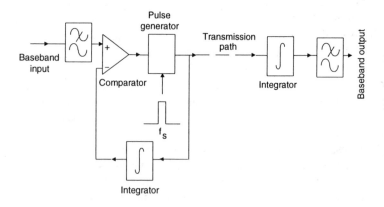

Figure 2.24   Basic delta modulation system

Alternatively, instead of transmitting a bipolar pulse train, a unipolar pulse train can be used (i.e. '1' = +V and '0' = zero volts). The integrators are then discharged at a constant rate, so that their output voltages decrease by one quantum step during a sample period not containing '1'.

As shown in Figure 2.24, a delta modulator is very simple and therefore cheap. It also has the useful property that all transmitted digits have equal weight, corresponding to one quantum step. Thus, a one bit error in transmission only causes an error of one quantum in the output signal.

In PCM, an error in the most significant digit causes an error of half the maximum signal amplitude, with disastrous results if this occurs frequently. Thus, DM is very robust in the presence of frequent transmission errors. It has therefore been used in military systems.

In theory, there is no limit to the amplitude of baseband signal which the system can transmit. It can handle a signal of any amplitude, providing that it is not changing so rapidly that it causes slope overload.

To avoid slope overload, the input signal, v_m, must not change by more than one quantum step (Δ) in the sampling interval, $1/f_s$, i.e. as in Equation 2.63. If v_m is given by Equation 2.64 then Equation 2.65 is obtained, its maximum value being given by Equation 2.66 so that V_m must satisfy Equation 2.67.

$$\frac{d\,v_m}{d\,t}\frac{1}{f_s} \leq \Delta \tag{2.63}$$

$$v_m = V_m \cos \omega_m t \tag{2.64}$$

$$\frac{d\,v_m}{d\,t} = -\omega_m V_m \sin \omega_m t \tag{2.65}$$

$$\omega_m V_m = 2\,\pi f_m V_m \tag{2.66}$$

$$V_m \leq \frac{\Delta}{2\,\pi}\frac{f_s}{f_m} \tag{2.67}$$

Thus, DM has a dynamic range which is large at low frequencies, but which decreases with frequency. This is satisfactory for speech, since speech signals contain little energy at high frequencies.

In DM, the output voltage increases and decreases in increments of one quantum step; it cannot remain stationary. The maximum quantising error is thus $\pm \Delta$ whereas in PCM it is $\pm \frac{1}{2}\Delta$. Consequently, the quantisation noise power is four times that given by Equation 2.48 i.e. as in Equation 2.68.

$$E^2 = \frac{1}{3}\Delta^2 \qquad (2.68)$$

However, the quantisation noise power spreads over a wider frequency band than for PCM, because of the higher sampling frequency (Wallace et al., 1991). If it is assumed that the noise power is uniformly distributed up to frequency f_s, then the fraction of the noise power reaching the output is F_m/f_s, where F_m is the cut off frequency of the baseband output channel, resulting in Equation 2.69.

$$N_o = \frac{\Delta^2}{3}\frac{F_m}{f_s} \qquad (2.69)$$

For a sinusoidal signal which just avoids slope overload, V_m is given by Equation 2.67 and the output signal power is given by Equation 2.70 and the SQNR by Equation 2.71.

$$S_o = \frac{1}{2}\left(\frac{\Delta}{2\pi}\frac{f_s}{f_m}\right)^2 \qquad (2.70)$$

$$SQNR = \frac{3}{8\pi^2}\frac{f_s^3}{f_m^2 F_m} \qquad (2.71)$$

For example, if $f_s = 64\text{kHz}$, $F_m = 3.4\text{kHz}$ and $f_m = 800\text{Hz}$, the SQNR is 36.6dB. This is similar to that with 64kbit/s PCM, but there is no companding to maintain the SQNR at lower signal levels.

2.6.2 Adaptive delta modulation

Simple DM is susceptible to slope overload. Moreover, since it provides uniform quantisation, the SQNR decreases when the level of the input signal is reduced.

Several schemes of adaptive delta modulation (ADM) have been introduced to minimise these disadvantages by adapting the step size under changing signal conditions (Steele, 1975; Peebles, 1976; Taub and Schilling, 1986).

In a basic DM system, a rapidly changing input signal results in a long sequence of 1's or 0's being transmitted. In continuously variable slope delta modulation (CVSDM), the rate of change of the integrator output is increased if there is a sequence of several 1's or 0's. Thus, the CVSDM coder can follow a rapidly changing signal more accurately (Taub and Schilling, 1976).

2.6.3 Delta sigma modulation

Because slope overload causes the dynamic range to decrease with frequency, simple DM is unsuitable for baseband signals having a flat frequency spectrum. This disadvantage can be overcome by preceding the delta modulator by an integrator to reduce the high frequency energy of its input signal, as shown in Figure 2.25(a). This is known as delta sigma modulation (DSM), (Steele, 1975; Peebles, 1976).

In order to obtain a flat overall gain/frequency response, a corresponding differentiator can be added at the output of the demodulator, as shown in Figure 2.25(a). However, the DM demodulator consists of an integrator, so differentiation of the output can be obtained by simply omitting it. The demodulator then consists only of a low pass filter, as shown in Figure 2.25(b).

Slope overload of the delta modulator within the delta sigma modulator occurs when dv_i/dt is too large. However v_i is given by Equation 2.72, resulting in Equation 2.73, and slope overload of this delta modulator corresponds to amplitude limiting of the input signal to the delta sigma modulator. Consequently, the signal amplitude that can be handled and the SQNR are both independent of frequency.

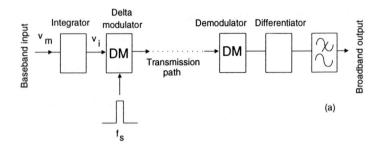

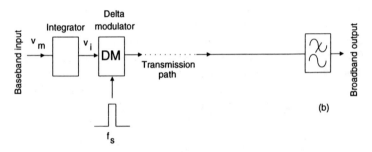

Figure 2.25 Delta sigma modulation: (a) derivation of delta sigma modulation; (b) delta sigma modulation system

$$v_i = k \int_0^t v_m \, dt \qquad (2.72)$$

$$\frac{d v_i}{d t} = k v_m \qquad (2.73)$$

For a sinusoidal baseband signal of maximum amplitude, it can be shown (Johnson, 1968) that, SQNR is given by Equation 2.74.

$$SQNR = \frac{9}{8 \pi^2} \left(\frac{f_s}{F_m} \right)^3 \qquad (2.74)$$

Thus, as with simple DM, the quantisation noise power decreases with the cube of the sampling frequency.

2.6.4 Conversion filters

The various digital modulation schemes can be represented by a general theoretical model (Flood and Hawksford, 1971), which applies time and amplitude quantising to an analogue modulated signal. For PCM, the analogue modulation is AM; for DM, it is PM; for DSM, it is FM.

It follows that one type of digital modulation can be converted to another. If DM is generated with a sampling frequency n times the Nyquist rate and the output of the delta modulator feeds an up-down binary counter, then output readings from the counter at intervals of n digits give PCM code words at the Nyquist rate. (Goodman, 1969). If DSM is used, the number of 1's in the digit stream should be counted over a period of n digits and the counter then reset (Wallace et al., 1991).

Counters used in this way are called conversion filters. Use of a delta modulator or a delta sigma modulator, together with a conversion filter, provides an alternative to A/D converters.

2.7 Coding programme quality sound

For broadcast programmes, the bandwidth and dynamic range required are greater than those needed for telephony and the permissible degradation is much smaller. The CCIR has set a standard of 32kHz sampling and 14-bit coding for sound channels of 14kHz bandwidth.

The ITU-T has recommended two alternative companding schemes: an 11-bit A-law, giving a reduction from 14 bits to 11 bits per sample, and a five range 'near instantaneous' companding scheme (NIC), giving a reduction from 14 bits to 10 bits per sample (Wallace et al., 1991).

2.8 Coding video signals

Composite video signals for colour television contain a luminance signal, together with colour components modulated onto quadrature sub carriers at 4.43MHz. Since the total bandwidth exceeds 5MHz, it might appear that the sampling frequency must exceed 10MHz. However, the spectrum about the colour sub carrier frequency is concentrated in lines at intervals equal to half the line repetition frequency. This enabled a lower sampling rate, f_s, to be chosen, such that the aliases from those lines above $f_s/2$ fall into gaps in the spectrum below $f_s/2$ and can be removed by comb filtering.

A sampling frequency of 8.86MHz and 8-bit encoding gives a digit rate of about 70Mbit/s. In practice, 68Mbit/s is used, but addition of error protection increases the digit rate further. The CCIR has recommended that 140Mbit/s transmission be used in order to provide a higher quality than that provided by analogue transmission (CCIR Report 962).

A number of techniques have been developed to reduce the bandwidth required for a 70Mbit/s video signal. Differential PCM can be used instead of conventional PCM. With non uniform quantisation, 4 to 6 bits are sufficient for the luminance signal and 2 to 4 bits for the colour information. This enables 34Mbit/s to be used for programme quality signals or 8Mbit/s for video telephony (CCIR Report 629-2).

More complex coding techniques have been developed to achieve further bandwidth reductions. (Jain, 1981). For example, the application of conditional replenishment and motion prediction enables the digit rate to be reduced to about 384kbit/s (Wallace et al., 1991). The performance obtained is adequate for colour video conferencing applications (ITU-T Recommendation H.261).

2.9 Acknowledgements

This chapter contains material reproduced by permission of Messrs. L.D. Humphrey, M.J. Sexton and A.D. Wallace and the Institution of Electrical Engineers from Chapter 6 of Flood, J.E. and Cochrane, P. (1991) Transmission Systems, Peter Peregrinus.

2.10 References

Alexander, A.A., Gryb, R.M. and Nast, D.W. (1960) Capabilities of the telephone network for data transmission, *Bell Syst. Tech. Jour.*, **39**, p.431.

Arthurs, E. and Dym, H. (1962) On the optimum detection of digital signals in the presence of white Gaussian noise, *IRE Trans.*, **CS-10**, pp. 336–373.

Atal, B.S. and Schroeder, M.R. (1970) Adaptive predictive encoding of speech signals, *Bell Syst. Tech. Jour.*, **49**, pp. 1973–1986.

Beckmann, P. (1967) *Probability in Communication Engineering,* Harcourt, Brace and World.

Bennett, W.R. and Davey, J.R. (1965) *Data transmission*, McGraw-Hill.

Betts, J.A. (1970) *Signal processing, modulation and noise,* English Universities Press.

Brewster, R.L. (Ed) (1986) *Data communication and networks,* Peter Peregrinus.

Brown, J. and Glazier, E.V.D. (1974) *Telecommunications*(2nd ed.), Chapman and Hall.

Bylanski, P. and Ingram, D.G.W. (1987) *Digital transmission systems* (3rd edn.), Peter Peregrinus.

Cattermole, K.W. (1969) *Principles of pulse code modulation*, Iliffe.

Chittleburgh, W.F.S., Green, D. and Heywood, A.W.A. (1957) A frequency-modulated voice-frequency telegraph system, *Post Off. Electr. Engrs. Jour.*, **50**, p. 69.

Clark, A.P. (1983) *Principles of digital data transmission* (2nd edn.), Pentech Press.

Clark, A.P. (1985) *Equalisers for digital modems,* Pentech Press.

de Belin, M. (1991) Microwave radio links, Chapter 12 in Flood, J.E. and Cochrane, P. (eds.) *Transmission Systems*, Peter Peregrinus.

Dorward, R.M. (1991) Digital transmission principles, Chapter 7 in *Transmission Systems*, Peter Peregrinus.

ETSI (1989) European Telecommunications Standards Institute *GSM full-rate speech transcoding,* GSM 06.10, version 3.01.02, April.

Evans, B.G. (Ed.) (1991) *Satellite communication systems* (2nd edn.), Peter Peregrinus.

Flood J.E. and Hawksford, M.J. (1971) An exact model for delta-modulation processes, *Proc. IEE*, **118**, pp. 1155–1161.

Goodman, D.J. (1969) The application of delta modulation to analogue-to-PCM encoding, *Bell Syst. Tech. Jour.*, **48**, pp. 321–343.

Harman, W.W. (1963) *Principles of the statistical theory of communication*, McGraw-Hill.

Hills, M.T. and Evans, B.G. (1973) *Transmission systems*, Allen and Unwin.

Jahnke, E. and Emde, F. (1960) *Tables of higher functions*, McGraw-Hill.

Jain, A.K. (1981) Image data compression: a review, *Proc. IEEE*, **69**, pp. 349–389.

Johnson, F.B. (1968) Calculating delta modulator performance, *IEEE Trans.*, **AU-16**, pp. 121–129.

Kretzmer, E.R. (1966) Generalisation of a technique for binary data transmission, *IEEE Trans.*, **COM-14**, pp. 67–68.

Lender, A. (1963) The duobinary technique for high speed data transmission, *AIEE Com. and Electron.*, **82**, pp. 214–218.

Lucky, R.W. and Hancock, J.C. (1962) On the optimum performance of N-ary systems having two degrees of freedom, *IRE Trans.*, **CS-10**, pp. 185–193.

Lucky, R.W., Salz, J. and Weldon, E.J. (1968) *Principles of data communication*, McGraw-Hill.

Nouri, A.M. (1991) *Satellite Communication*, Chapt. 13 in Flood, J.E. and Cochrane, P. (eds.) *Transmission Systems*, Peter Peregrinus.

Nyquist, H. (1928) Certain topics in telegraph transmission theory, *Trans. AIEE*, **47**, pp. 617–644.

Peebles, P.Z. (1976) *Communication system principles*, Addison-Wesley.

Purton, R.F. (1962) Survey of telephone speech-signal statistics and their significance in the choice of a PCM companding law, *Proc. IEE*, **109B**, pp. 60–66.

Ralphs, J.D. (1985) *Principles and practice of multi-frequency telegraphy*, Peter Peregrinus.

Sallen, R.P. and Key, E.L. (1955) A practical method of designing RC active filters, *IRE Trans.*, **CT-2**, pp. 74–85.

Saltz, J., Sheehan, J.R. and Paris, D.J. (1971) Data transmission by combined a.m. and p.m., *Bell Syst. Tech. Jour.*, **50**, pp. 2399–242.

Schwartz, M. (1970) *Information transmission, modulation and noise,* McGraw-Hill.

Schwartz, M., Bennett, W.R. and Stein, S. (1966) *Communication systems and techniques,* McGraw-Hill.

Sedra, A.S. and Smith, K.C. (1982) *Microelectronic circuits,* Holt, Rinehart and Winston.

Steele, R. (1975) *Delta modulation systems,* Pentech Press.

Sunde, E.D. (1959) Theoretical fundamentals of pulse transmission, *Bell Syst. Tech. Jour.*, **33**, pp. 721–788 and 987– 1010.

Taub, H. and Schilling, D.L. (1986) *Principles of communication systems,* (2nd edn.), McGraw-Hill.

Viterbi, A.J. (1966) *Principles of coherent communication,* McGraw-Hill.

Wallace, A.D., Humphrey, L.D. and Sexton, M.J. (1991) Analogue/digital conversion, Chapt.6 in Flood, J.E. and Cochrane, P. (eds.) *Transmission systems,* Peter Peregrinus.

Williams, M.B. The characteristics of telephone circuits in relation to data transmission, *Post Off. Electr. Engrs. Jour.* **59**, p.151.

3. Frequency division multiplexing

3.1 FDM principles

Frequency Division Multiplex (FDM) is the frequency translation of a number of individual standard telephony channels so that they can be stacked side by side and form a single wide band signal. This principle is illustrated in Figure 3.1.

Two identical telephony channels C_1 and C_2 (0.3 to 3.4kHz) are each mixed with a carrier frequency and combined. Only the lower sideband is selected from the mixing process and thus the channel frequencies are inverted on the combined signal. The spacing of the carrier dictates the spacing between the combined channels and is normally 4kHz i.e. $f_2 - f_1 = 4$kHz. The received path at B undergoes the opposite process.

As the carrier is not transmitted with the signal then f_1' and f_2' are separately generated at B. Any frequency difference between the carriers at A or B will impose a frequency shift on the recovered channel of $f_n - f_n'$.

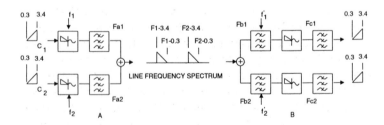

Figure 3.1 Frequency division multiplexing

The function of filters Fa1 and Fa2 is to select the wanted lower sideband and suppress the unwanted carrier signal, upper sideband etc. Fb1 and Fb2 filters select the required set of frequencies from the broadband line frequency spectrum for demodulation.

In the higher orders of translation it is not normally economic to design Fb filters with sufficient selectivity, such that only the required band of frequencies are selected.

The signal presented to the demodulator in this case can contain some information from the adjacent channels. These unwanted signals are removed by the greater selectivity of the lower order demodulating equipments.

An additional requirement for Fb filters is to limit the total band of frequencies applied to the demodulator and thus keep the intermodulation noise generated to a minimum. Fa and Fb filters are normally the same design.

The filters Fc1 and Fc2 remove the unwanted products from the demodulation process before presentation to the user or to the next lower order of translation.

The FDM system represents the most efficient use of bandwidth. With only the lower sideband of the modulation products transmitted the system limitations of bandwidth and overload are maximised.

3.2 History

The first UK FDM cable systems started service with a 12 channel system between Bristol and Plymouth in 1936 (Young, 1983). Previously all telephony transmission was at audio frequency and the move to FDM was inspired for economic reasons and the pressure on cable utilisation. From then onwards the drive was for lower loss and higher bandwidth cables to support ever higher bandwidth systems.

The period during the 1960s to the mid 1970s saw the peak of FDM systems in service and the development of transistorised systems to 60MHz. However during this period developments of digital systems started to fully occupy the R&D budgets and penetration of digital into the trunk transmission network rapidly overtook the FDM analogue network.

3.3 FDM hierarchy

3.3.1 General considerations

A number of different transmission media are available, such as open wire cable, coaxial cable, radio or satellite systems, all with different bandwidth capability. The multiplexing schemes use a hierarchy of building blocks to construct systems to the required bandwidth. Many stages of translation may be required with the final stages of modulation only being specific to the particular transmission medium.

The building blocks are optimised for cable transmission. The dimensions of the various blocks are largely historic and based on the economics of filter design.

3.3.2 Channel bandwidth

Channel spacing is standardised on 4kHz. This provides enough space between the voice frequencies (0.3 to 3.4kHz) to economically filter the carrier, pilots, outband signalling tones, etc. that are positioned in the gaps between the speech signals.

Where transmission bandwidth is at a premium then the channel spacing may alternatively be based on 3kHz (voice bandwidth 0.2 to 3.050kHz). This provides a 4/3 increase in channel capacity but is only achieved with increased cost of the channel translation stage.

3.3.3 Group and supergroup

With reference to Figure 3.2 the channel translating section converts 12 voice frequency channels (or 16 if 3kHz channelling) and assembles them into a basic 'group' in the range 60 to 108kHz.

Five groups are translated using carriers spaced 48kHz apart to form a 'supergroup' in the range 312kHz to 552kHz. A supergroup contains 60 channels.

Where a suffix has been added i.e. group 5 or supergroup 12, the suffix refers to the carrier that will be used to translate that particular

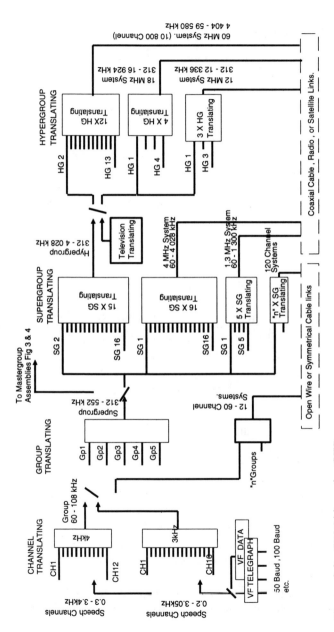

Figure 3.2 FDM hierarchy (UK)

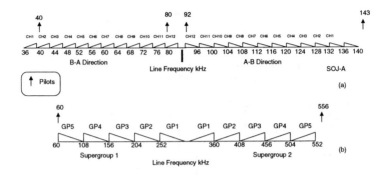

Figure 3.3   Typical open wire and symmetrical pair systems:
(a) open wire, 12 channel; (b) symmetrical pair, 120 channel

set of channels and identifies it in the higher order band of frequencies.

Groups and supergroups are used to construct systems on open wire and symmetrical cable. The line frequency spectrum of the typical systems are shown in Figure 3.3.

3.3.4   Higher order translation

Above supergroup level different administrations have adopted different hierarchies to build large systems. Three schemes have been identified and are described below:

1.   The 15 supergroup assemblies (UK).
2.   The mastergroup and supermastergroup assemblies (Europe).
3.   The Bell mastergroup plan (USA).

3.3.4.1 *15 supergroup assemblies*

Supergroups 2 to 16 (Figure 3.2) are translated to the band 312 to 4028kHz and form the 15 supergroup assembly or 'hypergroup'. Note

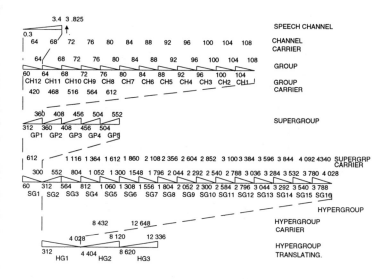

Figure 3.4   Translation frequencies and carriers (UK hierarchy)

that supergroup 2 is not translated and passed forward directly to sit in the higher order spectrum at 312 to 552kHz.

Hypergroups (900 channels) are used as building blocks to assemble large groupings up to 60MHz, the most important is the 12MHz 2700 channel system. The carrier frequency chart up to 12MHz is shown in Figure 3.4.

3.3.4.2 *Mastergroup*

The mastergroup (Figure 3.5) is a 5 supergroup assembly using supergroups 4 to 8 and spans 812kHz to 2044kHz. This is an alternative scheme for building large systems, which avoids the large 15 supergroup blocks but introduces more stage of translation.

Three mastergroups, mastergroup 7, 8 and 9, are translated to a supermastergroup 8516kHz to 12338kHz and supermastergroups are used in further translations stages for 12MHz, 18MHz, and 60MHz systems.

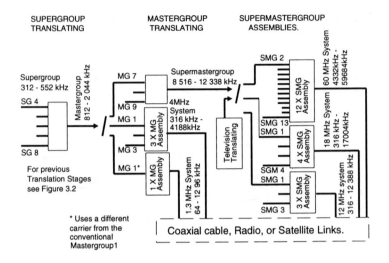

Figure 3.5 Mastergroup hierarchy

3.3.4.3 *Bell system*

The Bell system (Figure 3.6) has adopted a different mastergroup arrangement in that 10 supergroups are required to form a mastergroup. The supergroup carrier frequencies vary according to the system usage resulting in different mastergroup frequency allocations. The U600 mastergroup is used as a building block for further translations whereas the L600 mastergroup is used directly for transmission as a 600 channel system. Six mastergroups form a supermastergroup of 3600 channels.

3.4 Frequency translation

The ring bridge modulator/demodulator typically provides the general features for frequency translation. With reference to Figure 3.7, a sinusoid carrier is supplied at a high level (+10dBm to +13 dBm) sufficient to forward bias the diodes D_1 and D_2 or D_3 and D_4

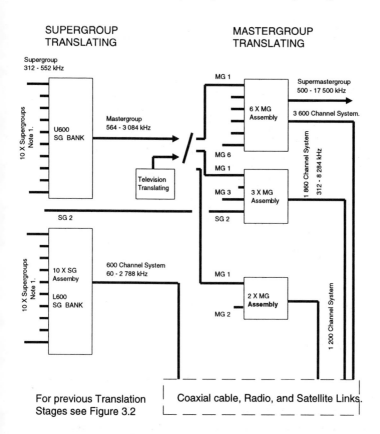

Figure 3.6   Bell FDM 'long haul' hierarchy. The Bell system Supergroup carrier frequencies do not always correspond to the CCITT (ITU-T) supergroup carrier frequencies

depending on the polarity of the carrier. The signal path, also through the diodes, is thus inverted on alternate half cycles of the carrier.

The current flowing, $i_c(t)$, due to the carrier signal is close to a square wave due to the clipping action of the diodes and by Fourier analysis is given by Equation 3.1, where k_1, k_3, k_5 etc. are circuit constants.

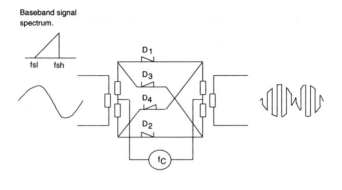

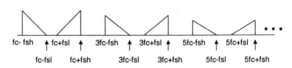

Figure 3.7 Ring bridge modulation

$$i_c(t) = k_1 \sin \omega_c t + k_3 \sin 3\omega_c t + k_5 \sin 5\omega_c t + \dots \quad (3.1)$$

If the input signal is represented by $i_{sig}(t)$ then the resulting modulated waveform at the output is given by Equation 3.2. i.e. the output frequency spectra is formed by the upper and lower sidebands about the carrier frequency and odd harmonics of the carrier frequency.

$$i_{out}(t) = i_c(t)\, i_{sig}(t)$$

$$= k_1\, i_{sig}(t) \sin \omega_c t$$

$$+ k_3\, i_{sig}(t) \sin 3\omega_c t + k_5\, i_{sig}(t) \sin 5\omega_c t + \dots \quad (3.2)$$

Note that neither the carrier frequency, carrier frequency harmonics, nor the original baseband signal is present at the output. i.e. the modulator is 'balanced'.

3.4.1 Ring bridge modulator/demodulator design considerations

3.4.1.1 *Carrier compression.*

With this type of modulator the ratio of the change in carrier power to the change in signal loss of the modulator, known as 'carrier compression', is approximately 10:1 so that accurate level stability of the carrier is not required.

3.4.1.2 *Carrier and signal suppression*

This is also known as carrier 'leak' and signal 'leak'. Perfect balance of modulator referred to above is not possible in practice. The best balance that can be achieved in a manufacturing environment is in the order of 30dB.

There will therefore be some unwanted products at the output of the modulator that will have to be removed by filtering before presentation to the combined path. Filter Fa (see Figure 3.1) requirements are shown in Table 3.1. The requirement is higher for the signal leak and the upper sideband as these products will give rise to crosstalk.

3.5 Carriers

3.5.1 Carrier frequency accuracy

For some services, particularly data circuits, it is desirable that the 'virtual' carrier difference over a network is maintained to within 2Hz (CCITT, 1985). To achieve this the carrier generation master oscillators receive regular maintenance adjustment using a reference frequency comparison pilot (see Section 3.6.2.2) or in later developed systems are phased locked to the frequency comparison pilot.

The recommended stability of the carrier frequencies are given in Table 3.2. With this carrier accuracy the end to end difference in virtual carriers will exceed 3Hz for 1% of the time and 4Hz for 0.1% of the time over a hypothetical international link of 2500km.

Table 3.1 Filter Fa requirements

Product	Level at modulator output	Requirement	Filter suppression
Carrier leak	+20dBmO	−50dBmo	70dB
Signal leak	−30dBmO	−80dBmO	50dB
Upper sideband	−30dBmO	−80dBmO	50dB

Table 3.2 Recommended stability of carrier frequencies

Carrier	Stability
Channel carriers	$\pm 1 \times 10^{-6}$
Group and supergroup	$\pm 1 \times 10^{-7}$
Hypergroup mastergroup and supermastergroup associated with 12MHz systems	$\pm 5 \times 10^{-8}$
Hypergroup mastergroup and supermastergroup associated with systems above 12MHz	$\pm 1 \times 10^{-8}$

3.5.2 Carrier purity

This is as follows:

1. Harmonics less than −20dBmO.
2. Purity against other carriers and side products related to 4kHz less than −80dBmO.

Spurious sidebands are produced by the carrier impurities during the modulation process and these create both intelligible and unintelligible crosstalk. The level of the 'ghost' sidebands is related to the level of the impurity of the carrier (Tucker, 1948).

3.5.3 Carrier level

Carriers are generated at a high level (+25dBm) for distribution to the translating equipment. Each equipment normally requires two supplies at a power level of approximately +10 to +13dBm in order to drive the modulator and demodulator.

3.6 Pilots

3.6.1 Translation equipment pilots

Reference pilots are discrete frequencies added to each translated band of traffic channels for monitoring and automatic gain control regulation at the far end of the transmission path. Some of the more common reference pilots are given in Table 3.3.

Additional pilots can be added for measurement or monitoring of wide band systems and are usually positioned in the gaps between supergroups. These are called intersupergroup pilots.

3.6.1.1 *Use of reference pilots for automatic gain control*

The normal practice is to provide gain or loss correction to the traffic path at the received end of transmission using the pilot to control the

Table 3.3   Some common reference pilots

Pilot	Frequency
Group reference pilot	84.080kHz
Supergroup reference pilot	411.920kHz
Hypergroup or mastergroup reference pilot	1552kHz

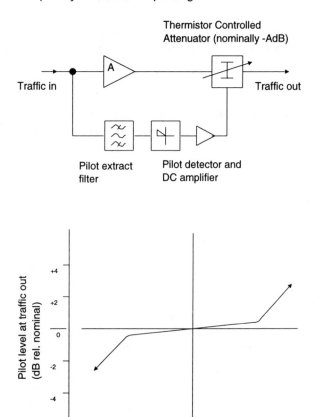

Figure 3.8 Automatic gain control (AGC)

correction. (See Figure 3.8). This correction is applied most commonly at group and supergroup level.

Correction is applied to maintain the design traffic level across the circuit link which may contain many stages of translation and several tandem connected line systems. Seasonal temperature changes to the cable link is foremost in the contributing factors. The correction is applied equally across the frequency band and with an Automatic Gain Control ratio of approximately 10:1.

Thermistors are normally used as gain control elements to avoid any unlinearity distortion to the traffic path. In addition the slow thermal time constant is suited to the slow diurnal gain change of the cable system and the envelope gain of the AGC relatively easy to control. ± 4dB regulation range is normally required from AGC equipments.

3.6.2 Line equipment pilots

Specific pilots are combined with the traffic path for transmission over copper line systems.

3.6.2.1 *Regulation pilots*

Regulating pilots are placed at the high end of cable frequency spectrum close to the top traffic frequency where the cable loss is greatest and the pilot most sensitive to any loss changes. The gain of the line repeater is modified to keep the pilot level constant throughout the seasonal changes of cable loss and thus the traffic power level at the repeater output can be maintained close to the thermal and intermodulation noise design optimum. The regulating pilot is transmitted at −10dBmO. (See also Section 3.14.3.)

Commonly used regulating pilot frequencies are given in Table 3.4.

3.6.2.2 *Frequency comparison pilots*

The frequency comparison pilot is used at each translation point to maintain the carrier generation master oscillators to a high degree of frequency stability. The pilots are placed at the low end of the cable frequency spectrum for minimum phase error shifts. These phase shifts occur from the cable and equalisers. Some commonly used frequency comparison pilots are given in Table 3.5.

Table 3.4 Commonly used regulating pilot frequencies

Coaxial system	Pilot frequency
1.3MHz, 300 channel	1364kHz
4MHz, 960 channel	4092kHz
4MHz, 900 channel	4287kHz
12MHz, 2700 channel	12435kHz
18MHz, 3600 channel	18480kHz
60MHz, 10800 channel	61160kHz

Table 3.5 Commonly used frequency comparison pilots

Coaxial system	Pilot frequency
1.3MHz, 300 channel	60kHz
4MHz, 960 channel	60kHz
4MHz, 900 channel	300kHz
12MHz, 2700 channel	308kHz
18MHz, 3600 channel	564kHz
60MHz, 10800 channel	564kHz

3.7 Noise contributions

Noise is the largest source of degradation to analogue systems.

Two contributions have to be considered, thermal and unlinearity noise. To evaluate the noise performance of an equipment both noise source contributions are calculated in pWOp and summated.

3.7.1 Definitions

There are three power levels in common use, dBr, dBm and dBmO.

3.7.1.1 *dBr*

The 0dBr point is the level at a reference point within the system that all the transmission levels refer. This used to be the 2 wire audio point or the 2 wire point of origin, as it was sometimes called.

This physical entity within a network has all but disappeared and a point of reference is now normally taken as the virtual outgoing switch point and is set at –4dBr.

From the transmissions point of view the audio output from the channel translating equipment is adjusted to suit the required stated dBr level at that point and this becomes the level to which all the transmission levels within the system refer. Transmission levels are referred to in dBr.

3.7.1.2 *dBm*

The power at a various points in the transmission are referred to in dBm. If a test signal is injected into a 0dBr point at 0dBm then the test signal level throughout the transmission can be referred to in dBm.

It is sometimes the case that for noise calculations where the transmission level is required in dBm rather than dBr it is assumed that the power at the 0dBr point is 0dBm.

3.7.1.3 *dBmO*

This is the power of a signal in dBm referred to a point where the transmission level is 0dBm.

For instance if a signal power is –80dBm at a 30dBm transmission level point, then the signal is defined as –50dBmO.

3.7.2 Psophometric weighting

When noise, either thermal or unlinear noise, is added to a telephone conversation, the degree of annoyance or the effect on the intelligibility of the conversation is not the same for all frequencies of added noise.

A weighting curve has been constructed to characterise this effect. Various curves exist (CCITT , Bell System C message etc.) all giving weighting factors of 2 to 2.5dB. The UK adopt the CCITT (now ITU-T) curve in Figure 3.9 providing a weighting factor of 2.5dB. This factor allows that for a 4kHz channel, considered over the speech bandwidth of 3.1kHz, 2.5dB more Gaussian type noise can be tolerated from the system when weighting is applied.

When this correction is applied the suffix 'p' is added to the noise power figure (i.e. dBmOp). An additional weighting factor is included when the full channel width of 4kHz is considered.

The weighting factor for a 4kHz channel considered over the full 4kHz is 2.5 + 10log(4/3.1) or 3.6dB.

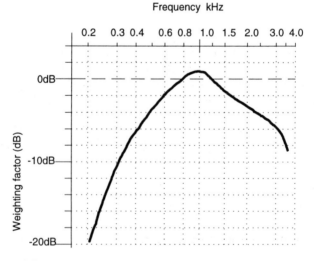

Figure 3.9   Psophometric weighting curve (ITU-T)

3.7.3 Thermal noise

This is sometimes called Browian or Gaussian noise and is noise that is associated with the random movement of electrons at temperatures above absolute zero.

For the FDM range of frequencies the noise power that is generated is given by Equation 3.3, where P is noise power in watts, k is Boltzmann's constant, T is the absolute temperature in degrees Kelvin, and B is bandwidth in Hz.

$$P = k\,T\,B \quad watts \tag{3.3}$$

If $T = 310^{\circ}K$ and $B = 4000Hz$ then the available noise power is -137.5dBm.

The absolute thermal noise power at the output of a circuit element of gain GdB and noise figure FdB within a 4kHz channel is given by Equation 3.4.

$$-137.5 + G + F \quad dBm \tag{3.4}$$

If the absolute power at the transmission level point T_L is known (TL in dBm) then the thermal noise contribution to the noise power in a 4kHz channel can be calculated in dBmOp (or pWattOp) as in expressions 3.5 and 3.6, where 3.6dB is the weighting factor for a 4kHz bandwidth. Definitions of the noise factor F are found in Connor, 1973.

$$-137.5 + G + F - T_L - 3.6 \quad dBmOp \tag{3.5}$$

$$\frac{1}{1000}\,Antilog\left(\frac{-137.5 + G + F - T_L - 3.6}{10}\right) \quad pWOp \tag{3.6}$$

3.7.4 Noise due to unlinearity

When carrying a traffic channel load the unlinear transfer characteristic of the active devices in the system generate harmonics and more complex intermodulation products that very quickly take on the characteristics of gaussian noise.

This noise source is referred to as intermodulation or cross modulation noise.

The initial problem is to define the power presented to the system by speech traffic.

3.7.4.1 *Single channel load*

The average power in each channel of a transmission band carrying normal 'busy hour' speech traffic referred to a 0dBr point of measurement is L_o (dBmO) as in Equation 3.7 (Bell, 1971), where P_{vo} is the long term average power in a single talker dBm0, g is the standard deviation of the distribution of talker volumes and t is the traffic activity factor.

$$L_o = P_{vo} + 0.115\, g^2 + 10 \log t \qquad dBmO \qquad (3.7)$$

Single talker volume (P_{vo})

This is the power in dBmO of a continuous individual talker measured using a long term averaging meter (10 seconds) at a 0dBr point. This includes power due to signalling frequencies.

The precise value for P_{vo} has been the subject of much measurement and study over the years. The UK figure used for trunk networks is –12.9dBmO. In the Bell System a figure of –13.9dBmO has been adopted for International use. (Most of the figures have been derived from Holbrook and Dixon, 1939.)

Loud and soft talker factor ($0.115g^2$)

P_{vo} is the average power in an individual talker. However, analysing a large number of talkers will show that they all have different volumes compared to the standard talker i.e. there are loud and soft talkers.

The distribution of the talker volumes (in dB) has been found to be Normal, with a standard deviation of g. (Figure 3.10(a).)

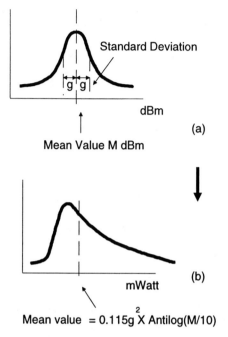

Figure 3.10   Log normal distribution

In order to find the average power per talker from a distribution of talker volumes the average or expected value is derived from the log normal distribution (see Figure 3.10(b)). This derives that the average power per talker over a large number of talkers is $0.115g^2$ greater than the average power of a single talker (Bennett, 1940).

Activity factor (10log t)

Not all the channels are fully occupied with continuous talkers. This gives rise to an activity factor t.

Conversation between two parties consist of 50% listening and 50% talking. The activity in any one direction is therefore 0.5.

In addition not all of the channels are in use. Even in the peak period only 70% of the circuits are occupied at any one time. The

Table 3.6   Values for L_o, $P_{vo,}$ g and t

	L_o (dBmO)	P_{vo} (dBmO)	g	t
CCITT (ITU-TS)	−15.0	−12.9	5.8	0.25
Bell International	−14.5	−13.9	5.0	0.45

activity factor t is therefore reduced to 0.35. The figure normally used for t is 0.25.

Values for L_o, P_{vo}, g and t are given in Table 3.6. (CCITT, 1985; Bell, 1971.)

3.7.4.2 *Multichannel load*

The average traffic power P_n in a multichannel load of N channels is given by Equation 3.8, where $N \geq 240$, assuming that a single channel load figure is given by the −15dBmO value in Table 3.6.

$$P_n = -15 + 10 \log N \quad dBmO \tag{3.8}$$

For N < 240 the statistics of the activity factor dictate that a slightly higher value for the average load should be used. A best fit curve is given by Equation 3.9 (CCITT, 1985).

$$P_n = -1 + 4 \log N \quad dBmO \tag{3.9}$$

3.7.4.3 *Unlinearity characterisation*

For a sinusoid fundamental signal at the output of a typical amplifier, the output power of the second and third harmonics would appear as in Figure 3.11. In order to calculate the effect of the unlinearity on the transmission it is necessary to be able to quantify the harmonics and associated intermodulation products.

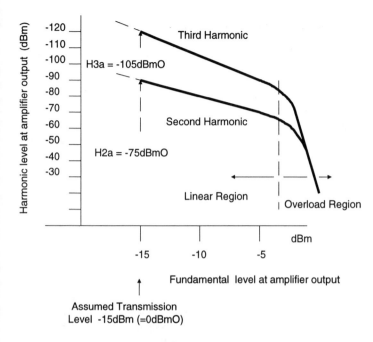

Figure 3.11   Amplifer characteristics

At the overload point, or clipping region, the harmonics increase very rapidly with increasing fundamental level and the characterisation of harmonic performance of the amplifier in this region is not practical.

As the output level is reduced and moves out of the overload region, the shape of the harmonic curve passes the knee and enters a linear portion of the graph where the second harmonic level reduces by 2dB and the third harmonic by 3db for every 1dB drop in fundamental signal. The harmonic performance is characterised in this region.

If the amplifier transfer characteristics in the 'linear' region are assumed to be a square law, Equation 3.10 is obtained, where v_i is the input voltage, given by Equation 3.11, v_o is the output voltage, and a_1, a_2, a_3 are constants of the amplifier.

$$v_0 = a_i v_i + a_2 v_i^2 + a_3 v_i^3 + \dots \tag{3.10}$$

$$V_i = V \cos \omega t \tag{3.11}$$

From this Equations 3.12 and 3.13 may be obtained.

$$v_o = a_1 V \cos \omega t \tag{3.12}$$

$$v_o = \frac{1}{2} a_2 V^2 + \left(a_1 V + \frac{3}{4} a_3 V^3 \right) \cos \omega t$$
$$+ \frac{1}{2} a_2 V^2 \cos 2 \omega t + \frac{1}{4} a_3 V^3 \cos 3 \omega t + \dots \tag{3.13}$$

Thus the second harmonic ($\cos 2 \omega t$) is proportional to the square of input signal amplitude V. Likewise the third harmonic ($\cos 3 \omega t$) is proportional to the cube of the input signal amplitude.

In dB terms it can be shown (Bell, 1971) that at a transmission level of 0dBmO the power in a sinusoidal harmonic or intermodulation tone, $P_m(x)$ in dBmO, at the amplifier output is related to the output level of the fundamental tones A and B by Equations 3.14 to 3.18, where $P_m(x)$ is the harmonic or intermodulation power in dBmO, P(x) is the fundamental power in dBmO, H_{2a} and H_{3a} are the second and third harmonic powers in dBmO, characterised for the amplifier with fundamentals at 0dBmO. (Figure 3.11.)

$$P_m(2A) = H_{2a} + 2P(A) \tag{3.14}$$

$$P_m(A \pm B) = H_{2a} + 6 + P(A) + P(B) \tag{3.15}$$

$$P_m(3A) = H_{3a} + 3P(A) \tag{3.16}$$

$$P_m(2A \pm B) = H_{3a} + 9.6 + 2P(A) + P(B) \tag{3.17}$$

$$P_m(A \pm B \pm C) = H_{3a} + 15.6$$
$$+ P(A) + P(B) + P(C) \tag{3.18}$$

Thus, once H_{2a} and H_{3a} have been determined by measurement, all the second and third order intermodulation products can be deduced for a specific transmission level.

3.7.4.4 *Determination of unlinearity noise from a multichannel load*

The technique used (Bennett, 1940) is to show that, for the purpose of calculating unlinearity or intermodulation noise, a channel loaded with a single sinusoid can be considered equivalent to a channel loaded with gaussian (or speech) noise by the application of a suitable factor $k(x)$ to the noise contribution . A band of n speech channels thus becomes a band of n sinusoids and the problem is reduced to one of counting intermodulation products falling into the channel of interest for each product $(A + B, 2A - B,$ etc.).

The total intermodulation noise in any particular channel is the summated power contribution from each of these products.

Bennett's formula, rearranged to give the weighted noise contribution $W(x)$ within a specified channel is as in Equation 3.19.

$$
\begin{aligned}
W(x) = {} & P_m(x) - k(x) + P_{vo}\,e(x) + 0.115\,g^2\,d(x) \\
& + 10 \log(U(x)\,t^{\mu(x)}) - C \quad dBmOp
\end{aligned}
\tag{3.19}
$$

The suffix (x) refers to the type of intermodulation under consideration i.e. $H(A + B)$ etc. (Table 3.7) In Equation 3.9 $P_m(x)$ is the power of the intermodulation product (x) in dBmO at the output of the system for 0dBmO fundamentals; $k(x)$ is the speech tone modulation factor (a factor in dB to convert the sinusoid $P_m(x)$ to the equivalent intermodulation product power for bands of 4kHz gaussian noise); P_{vo} is the power in a single average talker in dBmO (see Table 3.6); g is the standard deviation of the distribution of all talkers from loud to soft (see Table 3.6); $e(x)$ and $d(x)$ are factors to account for the relationship between the power in the talker (the fundamental signal) and the resulting intermodulation product $(e(x)$ is a factor to modify the talker volume P_{vo} and $d(x)$ to modify the standard deviation of talker volumes g); t is the transmission activity factor or the probability that a particular channel is active (see Table 3.6); $u(x)$ is the

Table 3.7 Types of intermodulation

Type(x)	k(x)	e(x)	d(x)	u(x)	C
A+B	0	2	2	2	3.6
A–B	0	2	2	2	3.6
A–2B	1.5	3	5	2	3.6
2A–B	1.5	3	5	2	3.6
2A+B	1.5	3	5	2	3.6
A+B+C	0	3	3	3	3.6
A+B–C	0	3	3	3	3.6
A–B–C	0	3	3	3	3.6

number of channels involved in forming the particular intermodulation product and therefore $t^{u(x)}$ is the probability that the particular intermodulation product from a particular set of channels is present; C is the psophometric weighting correction factor for 4kHz and is 3.6dB. U(x) is the number of intermodulation products for a particular type (i.e. A + B, A + B, etc.) falling in a particular channel of interest.

The factor U(x) is derived from Bennett's formulae but for simplicity are usually shown in graphical form as in Figure 3.12.

These graphs are valid for systems of greater than 500 channels (Bell, 1971).

3.7.4.5 *Approximate value for the weighted intermodulation noise contribution*

For wide band systems there are three major intermodulation contributors A + B, A – B and A + B – C. Assuming the ITU-T accepted values for channel loading (see Table 3.6) a rule of thumb calculation can be made for unlinearity contributions using Equations 3.20 to 3.22, where $W^*(x)$ is the approximate noise power in a selected

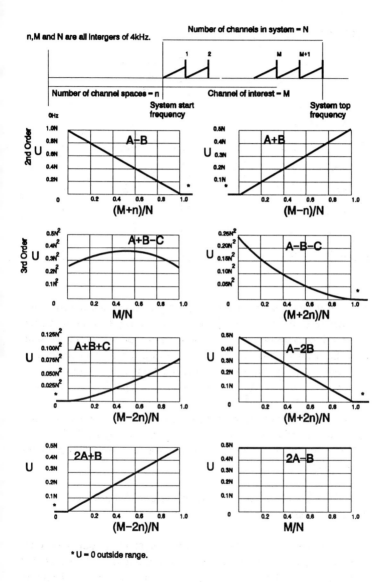

* U = 0 outside range.

Figure 3.12 U(x) values for large systems

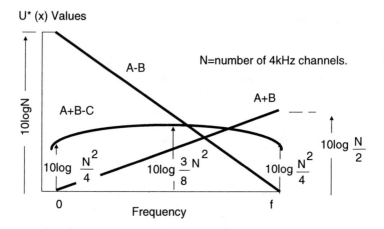

Figure 3.13   Approximate intermodulation factors U*(x)

channel, $P_m(x)$ is the power of the (x) product in dBmO for 0dBmO fundamentals, $U^*(x)$ is the factor obtained from Figure 3.13 for a channel at frequency f.

$$W^*(A+B) = P_m(A+B)$$
$$+ U^*(A+B) - 33.5 \quad dBmOp \tag{3.20}$$

$$W^*(A-B) = P_m(A-B)$$
$$+ U^*(A-B) - 33.5 \quad dBMOp \tag{3.21}$$

$$W^*(A+B-C) = P_m(A+B-C)$$
$$+ U^*(A+B-C) - 47 \quad dBmOp \tag{3.22}$$

3.7.4.6 *Weighted noise power in pWOp*

The total weighted intermodulation noise power can be determined from Equation 3.23, where x = all products, (A+B), (A−B), etc.

$$\sum \frac{1}{1000} \, Antilog \left(\frac{W(x)}{10} \right) \, pWOp \qquad (3.23)$$

3.7.4.7 Determination of unlinearity noise using spectral densities

The problem with Bennett's method is dealing with shaped frequency spectra (pre-emphasis at the output of Line repeaters for instance.)

A method using spectral densities (Bell, 1971) overcomes this difficulty and can be extended to include the shape of the H_{2a} and H_{3a} across the band.

From Equation 3.10 the ratio of the total second order distortion voltage to the fundamental signal at the output of the amplifier or system is given by Equation 3.24, where v_i is the system input signal (assumed Gaussian).

$$\frac{a_2 \, v_i^2}{a_1 \, v_i} \qquad (3.24)$$

The terms in Equation 3.24 can be expressed as a power in volts squared where $S_1(f)$ and $S_2(f)$ are the power spectral densities in volts squared per Hertz. $S_1(f)$ is the figure for the multichannel input signal voltage, v_i, and $S_2(f)$ for the input signal voltage squared, v_i^2. This is shown in Equation 3.25.

$$\frac{a_2^2 \, S_2(f)}{a_1^2 \, S_1(f)} \qquad (3.25)$$

If v_i is assumed Gaussian then Equation 3.26 may be obtained, ignoring d.c. terms and where * represents the convolution integral.

$$S_2(f) = 2 \, [\, S_1(f) \, * \, S_1(f) \,] \qquad (3.26)$$

The second order noise power NP_2 in watts is therefore given by Equation 3.27, where Pch is the traffic load per channel in watts.

$$NP2 = Pch \frac{a_2^2 \, 2 \, [\, S_1 \, (f) \, * \, S_1 \, (f) \,]}{a_1^2 \, S_1 \, (f)} \tag{3.27}$$

Likewise the third order noise power NP_3 in watts is given by Equation 3.28, where the mean square voltage is given by Equation 3.29 and Pch is the traffic load per channel in watts.

$$NP_3 = Psch \frac{a_3^2 \, [\, 9 \, V_{rms}^4 \, S_1 \, (f) + 6 \, (\, S_1 \, (f) \, * \, S_1 \, (f) \, * \, S_1 \, (f) \,) \,]}{a_1^2 \, S_1 \, (f)}$$

$$\tag{3.28}$$

$$V_{rms}^2 = \int_{-\infty}^{+\infty} S_1 \, (f) \, df \tag{3.29}$$

The convolution is best performed numerically on a per system basis.

3.8 Measurement of noise contributions

This is also referred to as white noise measurement. Both the thermal noise contributions and the intermodulation noise contributions can be measured directly with one test. This allows rapid field evaluation of installed systems and bench evaluation of individual circuits.

In Figure 3.14 the system is loaded with Gaussian noise from a generating source that has a flat frequency spectrum of noise, bandwidth limited by the filter F_{lim}. A single defined channel at frequency $f(c)$ is removed from the spectrum of noise by a filter F_{sc} before the noise band is transmitted into the system.

A measurement on the same channel at frequency $f(c)$ selected by filter F_{pc} at the receiver will therefore measure noise that has been generated only within the system itself.

The system is loaded with noise power P_s that represents the expected multichannel load, as in Equation 3.30, where P_s The total power of the signal applied to the system, P_n is the multichannel load for N channels, and T_L is the transmission level point in dBm.

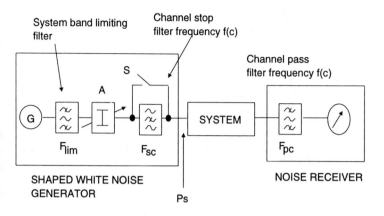

Figure 3.14   System loaded noise test

$$P_s = P_n + T_L \qquad (3.30)$$

For instance, the figures for a 4MHz system at a −30dBm transmission level point will be P_n = +14.8dBmO (assuming the normal channel loading of −15dBmO). P_s will evaluate to −15.2dBm at T_L= −30dBm.

Two measurements are taken. The first is with the channel stop filter at the transmitter bypassed with the switch S in the closed position. This gives the power due to simulated traffic in a single channel referred to the meter calibration. Measured figure = N_s.

The second measurement is with the channel stop filter at the transmitter in circuit with the switch S open. This gives the noise power generated from the system referred to the same meter calibration. Measured figure = N_n.

The Noise Power Ratio (NPR) is given by 10log(Ns/Nn) dB.

The total noise power falling into the channel can be determined from Equation 3.31, where N_c is the total weighted noise contribution over 4kHz in the specified channel in dBmOp, k = B/4N where B is the bandwidth of the system bandwidth limiting filter in kHz, and N is the number of channels.

$$N_c = -NPR - 18.6 - 10 \log k \quad dBmOp \qquad (3.31)$$

The term 10log k is to correct for the fact that the actual traffic load has gaps between the channels and is not continuous.

The total noise power is also given in pWOp by Equation 3.32.

$$\frac{1}{1000} Antilog \left(\frac{N_c}{10} \right) \quad pWOp \qquad (3.32)$$

Several channels are selected for measurement over the band by varying the frequency f(c) of both the channel stop filter in the transmitter and the channel pass filter in the receiver.

Also of interest is to vary the channel loading factor L_o (nominally −15dBmO) and determine the NPR for various conditions of traffic load. This is a measure of the system's robustness to peak loads and transmission level changes outside the normal design guides. The normal expected NPR curve is shown in Figure 3.15.

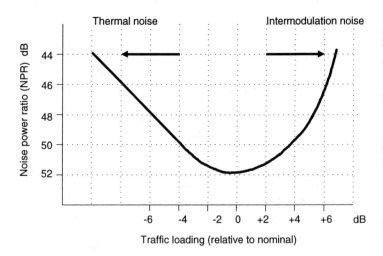

Figure 3.15 Noise power ratio variation for change in transmission loading

As the transmission load is lowered then the basic or Gaussian noise predominates. The NPR increases linearly with decreasing traffic power. As the transmission load is increased then intermodulation noise predominates and the NPR rapidly rises as more higher order intermodulation products contribute.

The optimum working point of the system is easily determined from Figure 3.15

3.9 Overload

Elements of the network have to be able to transmit the peak signal of the multichannel system load.

The system overload requirements are dimensioned such that the probability of the peak busy hour transmission signal exceeding the peak value of a sinusoid signal of power P_{eq} is 0.01.

P_{eq} in dBmO is given by Equation 3.33 (CCITT, 1985, Supplement 22) where t is the activity factor (see Table 3.6); L_0 is the channel loading factor (see Table 3.6); and n is the number of active channels.

$$
\begin{aligned}
P_{eq} = {} & L_o - 10 \log t \\
& + 10 \log \left[n + 2.33 \left(n \left(2 \, e^{0.23 \, g - 1} \right)^{\frac{1}{2}} \right) \right] \\
& + 9.9 + \frac{6}{1 + 0.07 \, n} \quad dBmO
\end{aligned}
\tag{3.33}
$$

The value of n is derived from assuming that the probability n channels are active in an N channel system is given by a binomial distribution and that n is only exceeded with a probability of 1% (i.e. the value of n is not exceeded 99% of the time) as in Equation 3.34. g is the standard deviation of loud/soft talkers (see Table 3.6) and N the system channel capacity.

$$
n = Nt + 2.33 \left(Nt \left(1 - t \right) \right)^{\frac{1}{2}}
\tag{3.34}
$$

Table 3.8 Examples of P_{eq}

Number of channels	12	120	960	2700
P_{eq} (dBmO)	19	21.2	27	30.5

P_{eq} has been calculated for some of the more important systems, as in Table 3.8.

3.9.1 Overload measurement.

The overload point is precisely defined according to one of two definitions, using harmonic/intermodulation products or using gain change.

3.9.1.1 *Harmonic/intermodulation products*

CCITT (1985) Recommendation G223, defines the overload as:

'The overload point or overload level of an amplifier is at that value of absolute power at the output at which the absolute power level of the third harmonic increases by 20dB when the input signal to the amplifier is increased by 1dB.'

If the 3rd harmonic is outside the frequency band of interest then the 2A–B intermodulation product may be used instead of the 3rd harmonic by the definition:

'The overload point or overload level of an amplifier is 6dB higher than the absolute power level in dBm, at the output of the amplifier, of each of two sinusoid signals of equal amplitude and of frequency A and B respectively, when these absolute power levels are so adjusted that an increase in 1dB of both of their separate levels at the input of the amplifier cause an increase, at the output of the amplifier, of 20dB in the intermodulation product of 2A–B.'

3.9.1.2 *Gain change*

From Equation 3.13 it can be seen the gain to the fundamental signal decreases by a factor given by expression 3.35 due to the non-linear characteristic.

$$20 \log \left(1 + \frac{3}{4} \frac{a_3}{a_1} V^2 \right) \tag{3.35}$$

This effect can be used to give a definition of overload such that:

'The overload is the output power in dBm of a signal such that a further increase in 1dB of a fundamental sinusoid signal power at the input of the amplifier will give an increase in power at the output of the amplifier of the fundamental signal of 0.9dB.'

The advantage of this last method is that it is easy to perform and can be measured at any frequency in the band of interest.

3.10 Hypothetical reference system

Hypothetical reference systems (CCITT, 1985, G.212 — G.222 & G.229) have been constructed to define network building and how the various degradation factors are apportioned and planned. The longest proposed international link is 25000km (CCITT,1985, G.103) and is constructed of a number of sub hypothetical links of typically 2500km.

3.10.1 Noise contributions

Noise contributions are defined on a per channel basis in pWOp. Recommended design targets are included in Tables 3.9 and 3.10.

The line equipment noise figure in Table 3.10 corresponds to 3pWOp per km. For wideband services, such as TV transmission, the equivalent requirement is more onerous at 1.5pWOp mean noise contribution per km.

Table 3.9 Design targets for noise contributions from open wire systems with 2500km circuit length

Equipment	Number of equipments	Total noise allocation
Channel translation	3	2500pWOp
Line (group) frequency translation	6	2500pWOp
Line equipment noise		17500pWOp

Table 3.10 Design targets for noise contirbutions from metric pair cable systems, coaxial cable and radio systems with 2500km circuit length

Equipment	Number of equipments	Total noise allocation
Channel translation 200pWOp	1	200pWOp
Group translation 80pWOp	3	240pWOp
Supergroup translation 60pWOp	6	360pWOp
Higher order translations 120pWOp	12	1440pWOp
Through connections etc.		260pWOp
Line equipment noise		7500pWOp

3.10.2 Line sections

The following applies:

1. The coaxial cable hypothetical systems are constructed from homogeneous line sections of 280km.
2. The change of gain of a regulated 280km homogeneous line section due to seasonal changes in the cable loss, time etc. will not be greater than 1dB.

3.   Unwanted modulation products from low frequencies (50Hz etc.) will not be greater than 48dBmO for a 280km homogeneous line section. A similar requirement is placed on the multiplex equipment.

3.11 Companding

Companding (compression/expansion) enables a noise advantage by compressing the dynamic range of the transmitted voice channel and performing the opposite expansion of the signal at the receiving end. Companders are used on long circuits where the noise additions have rendered the circuits unacceptable, or where a noisy transmission medium can be made viable.

3.11.1 Compander characteristics

The Compander parameters (Bell, 1971) are Compression Ratio and Unaffected Level, where compressor and expandor power output are given by Equations 3.36 and 3.37.

$$Compressor\ output\ (dBm) = \frac{Compressor\ power\ input\ (dBmO)}{K} + A \qquad (3.36)$$

$$Expandor\ power\ output\ (dBmO) = K\ (\ Expandor\ power\ input\ (dBmO)\ ) - K\,A \qquad (3.37)$$

K is normally 2 (the Compression Ratio) and KA equal 5 (the Unaffected Level). Compander characteristics are as in Figure 3.16.

The values for K and KA are limited by practical considerations. Increasing these values will improve the noise reduction but at the expense of other transmission properties. For instance increasing the compression ratio has the disadvantage of expanding any gain change or level error within the transmission.

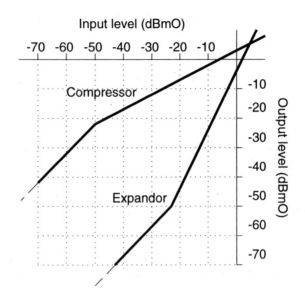

Figure 3.16 Idealised compressor and expandor characteristics

3.11.2 Multichannel load increase

Companding gives an increase in the peak multichannel load. The increase is mainly determined by the compression ratio, but the unaffected level and both the attack and decay times all impact on the overall loading.

Considering the main contribution from the compression ratio, the power in an single average talker is given in Table 3.6 where $P_{vo} = -12.9$dBmO, $g = 5.8$ and $t = 0.25$ to derive $L_o = -15$dBmO.

When compression is applied to the signal then the parameters above are modified by the compression ratio to give $P_{vo} = -12.9/2$dBmO, $g = 5.8/2$ and $t = 0.25$ to derive $L_o = -11.48$dBmO.

Using the new value for L_o (compressed) of -11.48dBmO new values for P_{eq} (compressed) can be calculated from Equation 30.24. For example a 4MHz 960 channel system has an overload require-

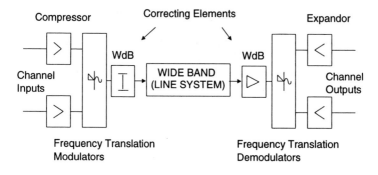

Figure 3.17   Transmission level correction for companding load increase

ment of +26.62dBmO with normal –15dBmO single channel loading. With compression the overload requirement is +29.73dBmO or 3.2dB higher.

It will therefore normally be necessary to drop the transmission level by 3.2dB (WdB) and to raise it again before expansion to avoid overloading the system (see Figure 3.17).

This adjustment to the transmission level will reduce some of the companding advantage by increasing the thermal noise contribution at the output of the expandor where the extra WdB gain is applied.

3.11.3 Compandor noise advantage

From Figure 3.16 it can be seen that while no speech is transmitted the noise power will be subject to an attenuation of 25dB at the expandor output.

The theoretical noise advantage will therefore be given by Equation 3.38.

$$Noise\ advantage\ =\ 25 - W\ \ dB \tag{3.38}$$

However, the circuit noise while speech power is transmitted is virtually unaffected by the compression/expansion process and there is no advantage. During talking, therefore, the conversation is subject to the full effect of the system noise. The effect of this noise produces an impairment to speech intelligibility that detracts from the companding advantage by an estimated 5dB.

Overall advantage is therefore given by Equation 3.39.

$$Noise\ advantage\ =\ 25 - W - 5\ \ dB \tag{3.39}$$

3.11.4 Attack and decay time

The compression and expansion process is speech power activated and it therefore takes time to recognise the increase (attack time) or decrease (recovery time) in speech power. Attack time is normally set at 5ms and recovery time 22.3ms. (CCITT, 1985, Recommendation G.162.)

3.11.5 Usage of companders

Companders are recommended for use with speech transmission if the mean noise power of a circuit in any hour is greater than 40000pWOp (–44dBmOp). (CCITT, 1985, Recommendation G.143.) The comparable limit for telegraphy is 80000pWOp (–41dBmOp). (CCITT, 1985, Recommendation H.21.)

3.12 Through connections

Within the network it may be required that a band of channels (typically group or supergroup) are allocated as a through connection path without translation to voice band. This could be at a drop and insert point for instance as shown in Figure 3.18.

3.12.1 Through connection filter

In this situation the band of interest has to be cleaned before passed back into the modulation process. From Figure 3.1 and Figure 3.19 a typical demodulator filter F_b selects the required band of frequencies

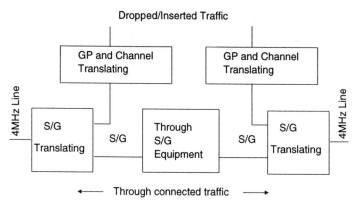

Figure 3.18   Drop and insertion of traffic

(band N) from the high order broadband of channels. Because of the selectivity of the filter F_b, the selected band of frequencies contain some of the adjacent channels from N+1 and N−1.

Normally these are removed by subsequent filtering and demodulation to voice level. If presented for re-translation however the adjacent channels must be removed to prevent crosstalk. Suppression of all such possible crosstalk paths is less than 80dBmO.

Such filters are large and complex and are usually separate equipments.

3.13 Transmultiplexers

A transmultiplexer provides a link between FDM and TDM hierarchies without having to translate to baseband voice frequencies.

The drive for transmultiplexer development came from analogue network operators updating to digital switches. With a digital switch the interface to the trunk network is at a 2.048kbit/s CEPT defined digital bit stream (or 1.554Mbit/s T1 for the USA digital hierarchy) i.e. there is no point that is suitable for direct connection into an FDM hierarchy without the cost and attendant distortions of translation to speech band.

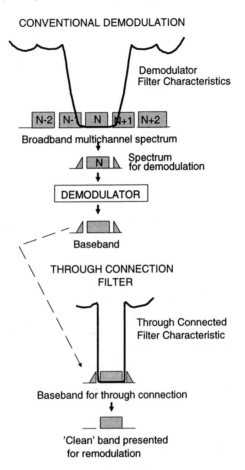

CONVENTIONAL DEMODULATION

Demodulator
Filter Characteristics

N-2 N-1 N N+1 N+2

Broadband multichannel spectrum

N Spectrum
for demodulation

DEMODULATOR

Baseband

THROUGH CONNECTION
FILTER

Through Connected
Filter Characteristic

Baseband for through connection

'Clean' band presented
for remodulation

Figure 3.19 Through connection filters

As a prime function therefore transmultiplexers interface to
2048kbit/s 30 channel (1544kbit/s 24 channel).

An economic fit of this digital interface is normally achieved by
taking 2 × 2048kbit/s 30 channel bit streams with a total of 60
channels and transmultiplexing to a single FDM supergroup.

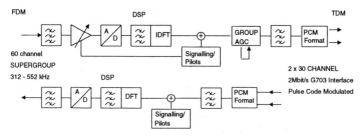

Figure 3.20   Supergroup transmultiplexer

The alternatively fit for a T1 1544kbit/s 24 channel bit stream is transmultiplexed into $2 \times$ FDM groups each carrying 12 channels. A block diagram of a typical 60 channel Transmultiplexer is shown in Figure 3.20 (Rossiter et. al.,1982).

The incoming supergroup is filtered to remove the adjacent sidebands. The gain is adjusted using an AGC from the supergroup pilot extracted within the Discrete Fourier Transform (DFT) and the complete band digitised through an analogue to digital converter. Using Digital Signal Processing (DSP) techniques the digitised band is split and then transformed into a TDM representation.

At this point the coefficients representing the pilots and signalling are extracted. The signalling is processed separately and the process algorithms depend on the type of signalling adopted (i.e. CCITT R1 or R2, E&M scheme etc.). The TDM band is then formatted for A-law companding, PCM coding and time processed for the 2048kbit/s CEPT structure.

The opposite procedure is adopted for the PCM to FDM conversion.

3.13.1 Synchronisation

The frequency accuracy requirement of an FDM supergroup is better than 1 part in 10^7 whereas the TDM stream is 50 part in 10^6 (Ribeyre et. al., 1982). To achieve the FDM accuracy the transmultiplexer may have to be locked to an accurate reference and allowance made to frame slip the TDM data.

3.13.2 PCM alarms

Data and alarms carried on TS0 of the PCM frame, data services carried on either the TDM or FDM signals and common channel signalling schemes have to be terminated at the transmultiplexer and treated separately.

3.14 Repeatered cable line equipment

The main function of the line terminal equipment is to service the cable transmission equipment. A supervisory overlay system is usually provided whereby a faulty repeater can be located from the terminal end. A constant current d.c. supply is established on the copper transmission wires for supplying power to the dependent repeaters (usually 50 or 110 mA).

Dependent repeaters are installed at regular intervals along the cable, the recommended spacing is as in Table 3.11. Repeaters are normally housed in pressurised boxes, either pole mounted or buried depending on the siting of the cable.

A coaxial line is required to meet an average noise contribution per circuit of better then 1.5pWOp per km. In order to achieve this with a limited DC power budget optimal designs have to be used.

Table 3.11   Repeatered cable spans

System	Cable	Repeatered span
1.3MHz (300 channels)	1.2/2.4mm	8km
4MHz (960 channels)	1.2/2.4mm	4km
12MHz (2700 channels)	1.2/2.4mm	2km
18MHz (3600 channels)	2.6/9.5mm	4km
60MHz (10800 channels)	2.6/9.5mm	1.5km

3.14.1 Pre-Emphasis

The noise contributions from a single repeater are shown in Figure 3.21(a). The greatest contribution is thermal noise at high frequencies as the highest gain is required at this end to overcome to cable loss.

By applying deliberate shaping to the transmission path at the transmitting terminal end (pre-emphasis) and an opposite characteristic at the receiving terminal end (de-emphasis) the noise contributions can be flattened so that all the channels have a similar contribution and the worst channels are 3 to 4dB improved.

In other words, the shape of the pre-emphasis balances the multi-channel load and thermal noise contributions so that these are equally

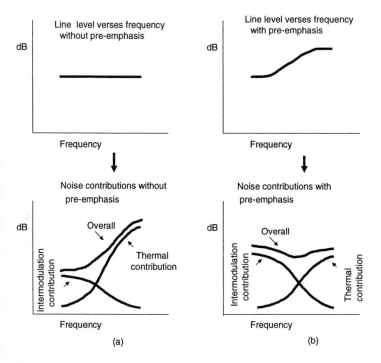

Figure 3.21   Principles of pre-emphasis

distributed over the line spectrum and all channels have ideally the same performance as Figure 3.21(b).

3.14.2 Thermal noise

To optimise the noise figure of the repeater and also to provide a good input impedance matching to the line produces conflicting requirements at the amplifier input. A good input impedance match to the cable is required with reflection losses normally kept to below 55dB and at the same time effective noise impedance matching is essential to give a low noise figure F.

Use of a Hybrid circuit at the input as shown in Figure 3.22 (Bell, 1971) provides a method of achieving both these requirements. The noise matching is improved by about 3dB compared to a conventional amplifier.

This technique can be extended further by including a hybrid transformer at the output and bridging the feedback between the input and output hybrid transformers.

3.14.3 Regulation

The gain shaping of a repeater is designed to match the loss characteristics of the cable. This is characterised by Equation 3.40, the loss being in dB.

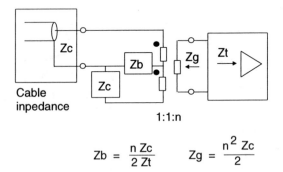

$$Zb = \frac{n\,Zc}{2\,Zt} \qquad Zg = \frac{n^2\,Zc}{2}$$

Figure 3.22   Hybrid input circuit

$$Loss\ at\ frequency\ f = Loss\ at\ frequency\ f_r \left(\frac{f}{f_r} \right)^{\frac{1}{2}} \qquad (3.40)$$

Thus if the loss of the cable is known at one frequency f_r (normally the regulating pilot frequency) the loss at other frequencies can be deduced.

The change of the cable loss with temperature obeys the same law and thus correction for temperature is applied with a root f shape across the line frequency spectrum with the greatest correction at the highest frequency.

This equalisation which is controlled by the regulating pilot is not exact and small errors accumulate in the gain frequency response at each repeater equalisation point.

The error systematically adds along the repeatered cable and to avoid excessive systematic error build up, a requirement to maintain the error to within 1dB at any frequency over a 280km system is necessary.

3.14.3.1 *Regulation range*

The change in loss of typical coaxial cable is approximately 0.004% per ^{o}C. The change in buried cable temperature in the UK is ±10°C.

Thus for a typical cable section loss of 40dB at the regulating pilot frequency the maximum diurnal change is approximately ±1.6dB. Over two such sections the loss change would be 3.2dB and with reference to Figure 3.15 it can be seen that a change in transmission load of 3.2dB produces only a small increase in NPR.

Thus regulation points in the UK are established every two repeatered sections as in Figure 23(a) and allow for a loss change of ±4dB. A higher temperature shift would require regulation at every repeatered point with the greater cost and power feeding requirements this would entail.

An alternative method of regulation, applying pre-regulation to halve the diurnal changes in repeater levels, is shown in Figure 3.23(b).

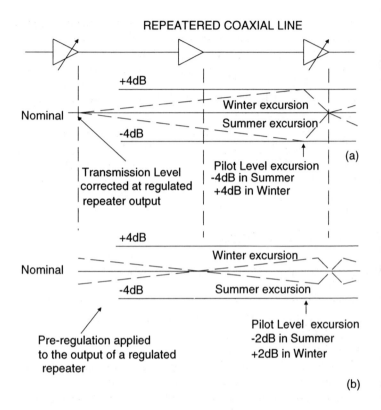

Figure 3.23 Repeatered line regulation: (a) pilot levels using conventional regulation; (b) pilot levels using pre-regulation

3.14.4 Power feeding

A constant current is normally used to serially feed the repeatered cable. A constant current ensures a consistent power delivered to each repeater even though the cable resistance is continuously shifting with the cable temperature.

The power is normally fed from both ends of the section with a turn round break in the centre. This break is important as the induction of unwanted currents from services (rail traction, grid systems etc.) that may run parallel with the transmission is proportional to the square of the effective coupled path (CCITT, 1985, K17).

Under traction or grid faults where very large currents may flow for a short period in the service cable, excessive voltages may be coupled into the transmission cable and can damage the repeaters and cause an insulation breakdown within the cable.

Even under normal conditions a long coupled path with a.c. traction or similar installations can cause the transmission power feed current to be modulated with the traction frequency, or its harmonics. This will create unwanted modulation products within the required multichannel traffic load.

3.15 References

Bell (1971) *Transmission systems for communications*, Bell Telephone Laboratories.

Bennett, W.R. (1940) Cross modulation requirements on multichannel amplifiers below overload, *Bell Systems Technical Journal*, (19) October.

CCITT (1985) Volume III, CCITT, Geneva.

Connor, F.R. (1973) *Introductory topics in electronics and telecommunications — Noise*, Edward Arnold, London.

Holbrook, B.D. and Dixon, J.T. (1939) Load rating theory for multichannel amplifiers, *Bell Systems Technical Journal*, (18), October.

Ribeyre, M. et. al. (1982) Exploration of transmultiplexers in telecommunication networks, *IEEE Trans. on Commun., COM-30, July, pp. 1493-1497.*

Rossiter, T.J.M. et. al. (1982) A modular transmultiplexer system using custom LSI devices, *IEEE Trans. on Commun.*, **COM-30**, (7), July.

Tucker, D.C. (1948) *Journal Inst. Electrical Engineers*, Part III, May.

Young, P. (1983) *The power of speech*, George Allen, London.

4. Time division multiplexing

4.1 General definition

In telephone systems a time division multiplexer, better known as a TDM, is generally defined as a device that distributes a number of channels periodically in time through the intermediary of pulse modulation. Each pulse corresponds to a channel and is interleaved between those of other channels. Hence, a time division multiplexed signal is always composed by means of synchronous sampling of the channels, with pulses shifted with respect to each other. This is illustrated in Figure 4.1. The interleaved channels form one frame of a duration corresponding to the sampling period T_s. The pulse modulation used can be analog (e.g. PAM, PPM) or digital (PCM).

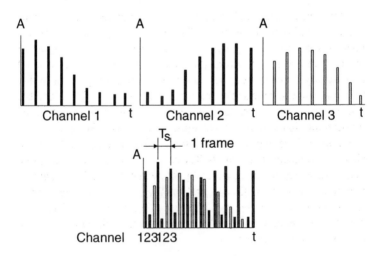

Figure 4.1   Time division multiplexing

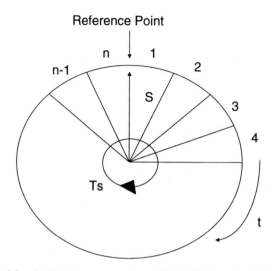

Figure 4.2   Cyclical representation of time division multiplexing

With time division multiplexing time is a relative value, i.e. TDM requires a reference point in the frame cycle to which the receiver must be synchronised in order to correctly demultiplex the stream of pulses it receives and extract the signals concerning each channel individually. Since the receiver must be continuously synchronised in both frequency and phase this reference point is repeated periodically. The easiest way to understand this is to cyclically represent the time division multiplexing process as illustrated in Figure 4.2.

Figure 4.2 depicts a rotating switch S scanning a number of channels connected to its inputs. The channels are scanned one by one in the order they are encountered by the switch: channel 1 is sampled at time t_1, channel 2 at time t_2, and finally channel n at time t_n, after which the process is repeated. The resulting output signal is composed of the samples of the different channels, shifted over a time period given by Equation 4.1.

$$\Delta t = t_n - t_{n-1} \tag{4.1}$$

If the channels presented at the input are pulse amplitude modulated signals, we obtain a signal as depicted in Figure 4.1. At the other side of the connection we find an identical switch executing the opposite operation, i.e. scanning the composite input signal at exactly the same frequency and phase, it demultiplexes the input signal and distributes the extracted signals over the respective channels. Both systems synchronise each time they pass the reference point.

Time division multiplexing has become very important, not only in transmission but also in switching, particularly in connection to the digital systems.

In the early days of electrical communications a medium such as copper wire carried a single information channel. For economic reasons, in terms of both cost and equipment, it was necessary to find ways of packing multiple channels onto one physical link. The resulting system is referred to as a carrier. Digital signals are now transmitted from one location to another by transmission facilities or systems using a multitude of media (paired cable, coaxial cable, analog or digital radio systems, optical fibres, satellite communication).

The synchronous time division multiplexers typically are used as termination equipment of such carrier systems. These multiplexers are better known as T1 or E1 multiplexers, depending on the multiplexing scheme they use. T1 is generally used in America and is based on the PCM system that was originally designed by AT&T. The name T1 was derived from the identification number of the committee set up by the American National Standards Institute (ANSI). Europe, on the other hand, uses the E1 system, which is based on CEPT (Conference of Post and Telecommunications) recommendations. The delay with which CEPT has undertaken the definition of a primary digital PCM system has allowed it to profit from the experience of the American systems.

On the international level, the two types of system coexist although they are incompatible, and are a subject of two recommendations of the CCITT (now the ITU-T). Although both systems will be discussed later in this chapter, Table 4.1 compares their main characteristics. The meaning of the different items will become clear throughout the rest of this chapter.

Table 4.1   Primary European and American systems

	E1 (European System) ITU-TS Recommendation G.732	T1 (American System) ITU-TS Recommendation G.733
Sampling Frequency	$f_s = 8\text{kHz}$	$f_s = 8\text{kHz}$
Bit rate per channel	$D_i = 64\text{kbit/s}$	$D_i = 64\text{kbit/s}$
Number of time slots	32	24
Number of channels	30	24
Number of bits/frame	$32 \times 8 = 256$	$24 \times 8 + 1 = 193$
Total bit rate	$256 \times 8\text{kHz} =$ 2.048kbit/s grouped word of 7 bits in the 0 channel of odd frames	$193 \times 8\text{kHz} =$ 1.544kbit/s distributed sequence 101010... consisting of the 193rd bit of odd frames
Signalling	Out-of-octet, grouped in channel 16, consisting of 4 bits per channel, distributed over 16 frames (=1 multiframe)	

Before discussing the T1 and E1 technologies in detail, consider first the general structure of a digital time division multiplex.

4.2 Digital TDM structure

4.2.1 Frame organisation

For each digital channel there is assigned a group of b bits, called a word which corresponds to a digital message element (data). The groups belonging to the same channel are transmitted at a frequency f_s, equal to the sampling frequency. Thus, the bit rate of channel i can be calculated as in Equation 4.2.

$$D_i = f_s b \tag{4.2}$$

In the case of digital PCM telephony, these words are composed of 8 bits and called octets. Hence the bit rate of each channel equals 64000bit/s.

When n digital channels are assembled into a time division multiplex, the collection of the n words of b bits (and eventually auxiliary bits which are added to them), within a period given by Equation 4.3, constitutes a frame.

$$T_s = \frac{1}{f_s} \tag{4.3}$$

Although the structure of the frame is strictly repetitive, their content obviously is not because the channels contain variable digital information. Two types of frame organisation can be considered, as shown in Figure 4.3:

1. Word interleaved, where the interleaving is performed on a character by character basis, i.e. a character from each data source is accumulated and combined as a word or frame for transmission. To do so, the frame is subdivided into $y \geq z$ time slots each containing b grouped bits corresponding to the same digital channel or auxiliary bits.
2. Bit interleaved, where the interleaving is performed on a bit by bit basis, i.e. one bit from each data source is accumulated

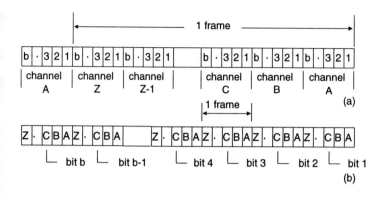

Figure 4.3 Digital TDM frame structure

and combined as a word or frame for transmission. To do so, the frame is subdivided into b groups each containing z bits of the same order belonging to each of the channels.

The word interleaved structure corresponds well to the mode of functioning of a PCM modulator, while the bit interleaved structure has advantages for time division digital switching because it suffers less throughput delay.

4.2.2 Frame alignment

The frame alignment or framing is typical of time division multiplexing. It consists of synchronising the receiving equipment both in frequency and in phase to the stream of symbols it receives. This operation is obviously necessary each time the receiver is switched on, but also during normal operation. Indeed, once aligned, the receiver needs a periodic time reference in order to check its isochronism and detect eventual shifts.

This necessary time reference consists of a particular pattern of several bits, called the framing pattern. When the receiver has lost frame alignment, it searches for this pattern in order to realign itself with as short a delay as possible. The framing pattern is periodically carried by the frame according to one of the following organisations:

1.   Grouped framing pattern, which consists of a number of consecutive bits at the beginning of a frame.
2.   Distributed framing pattern which, as the name suggests, is spread over the frame on a bit by bit basis, or over several frames at one bit per frame.

There is of course a danger of simulating the framing pattern by chance combination of other information carrying bits. There are different ways to protect the system from this, such as:

1.   One can choose a framing pattern with low autocorrelation, so that it is impossible to imitate by shifting and infringement on random neighbouring bits. Examples are the patterns 110 or 1110010 for for 3-bits or 7-bits grouped framing patterns respectively.
2.   One can block all the channels of the frame when the framing is lost at reception. The frame would be replaced by the transmission of a resynchronisation signal. This necessitates an announcement warning in the opposite direction, which is assumed to be correctly aligned.
3.   One could confirm correct framing by a criterion other than the presence of the framing pattern, e.g. absence of the pattern in one frame out of two.

In order to avoid reacting to each transmission error, the reaction to an incorrect reception of the framing pattern in an aligned situation, i.e. during normal data transfer, must be delayed. The solution to the framing problems of both T1 and E1 are given in the respective sections later.

4.2.3 Signalling

The transmission of multiple telephone calls is not the only requirement for a T1 or E1 network. The network also has to transfer signalling information. Signalling information is auxiliary information (digits, commands, acknowledgement signals, etc.) sent from exchange to exchange to control switching and management operation of the network.

Aside from analog signalling by sinusoidal carriers (in-band or out-band) used in analog systems and translatable by PCM modulation, digital systems lend themselves by nature much better to direct digital transmission of signalling information. Several solutions can be considered:

1.   In-octet signalling, which is also called bit robbing. In this the least significant bit (LSB) among the 8 bits that represent a coded sample of speech is periodically (typically every six frames) assigned to signalling. This results in an imperceptible degradation of the telephone transmission. However, when data are transferred over the same network, the impact of losing the LSB every sixth byte is not acceptable.

2.   Out-of-octet signalling, also referred to as channel by channel signalling. In this system each digital channel arranges, besides its octet, one or several signalling bits. These signalling bits of the n channels of the frame can be distributed, i.e. juxtaposed at each octet as a supplementary bit per time slot (possible signalling rate being 8kbit/s per channel), or grouped in a time slot reserved for this purpose and of which the bits are assigned in turn, cyclically, to the n channels of the frame (possible signalling rate being 64/n kbit/s per channel)

3.   Common signalling, where one time slot per frame is reserved for signalling and assigned according to need to one channel, then to another. Signalling is done using labelled messages, of which the label indicates which channel the message belongs to. This technique allows an instantaneous signalling rate of 64kbit/s for one channel at a time.

4.3 The digital hierarchy levels

A range of digital systems with increasing capacity has been defined, the systems of each order being composed of four systems of the immediately lower order. In North America the traditional TDM hierarchy is described as DS levels 0 through 4. Europe simply uses digits to indicate the digital order. These digital streams, produced by multiplexing equipment, are by design independent of the target transmission medium. In fact, in an end to end circuit, many different types of media may be encountered.

The 0kHz to 4kHz nominal voice band channels are first converted to digital information by PCM techniques and then stacked (multiplexed) onto higher bit streams. Each of the individual digitised 64kbit/s channels is referred to as DS0 levels (USA) or 0 order systems (Europe). Table 4.2 summarises the American and European digital hierarchies. In the USA 24 voice band analog channels are combined or multiplexed to form a DS1 signal (1.544Mbit/s), also called a digroup (for digital group). The rest of the digital hierarchy uses 3.152Mbit/s for 48 channels (DS1-C), 6.312Mbit/s for 96 channels (DS2), 44.736Mbit/s for 672 channels (DS3), and 274.176Mbit/s for 4032 channels (DS4).

The European telephone system is based on 30-channel blocks and uses transmission rates of 2.048Mbit/s to carry 30 channels (1st order), 8.448Mbit/s to carry 120 channels (2nd order), 34.368Mbit/s

Table 4.2 North American and European digital hierarchy

Order		Number of telephone channels at 64kbit/s		Total bit rate (Mbit/s)	
Europe	USA	Europe	USA	Europe	USA
0	DSO	1	1	0.064	0.064
1	DS1	30	24	2.048	1.544
2	DS1-C	120	48	8.448	3.152
3	DS2	480	96	34.368	6.312
4	DS3	1920	672	139.264	44.736
5	DS4	7680	4032	565.148	274.176

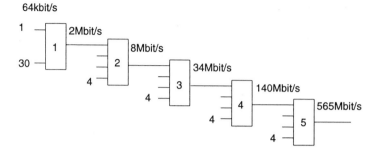

Figure 4.4   Hierarchy of digital systems

for 480 channels (3rd order), 139.264Mbit/s for 1920 channels (4th order), and 565.148Mbit/s for 7680 channels (5th order).

Figure 4.4 illustrates the coupling of the hierarchical systems. The high order services are structured around a point to point plesiochronous sectioned concept. This concept has a layered approach in which the 2Mbit data stream is supported by a 8Mbit server network. In turn this is supported by a 34Mbit server network which is itself supported by a 140Mbit server network. Starting from the second order, the equipment no longer contains an analog to digital converter and only deals with digital frames. The equipment essentially consists of multiplexers, combining into new frames the frames of four systems of the preceding order, called tributaries.

During the construction of an nth order multiplexer starting from frames with order n-1, we are confronted with the problem of aniso-chronism of the tributaries. In effect, the frames to be grouped come from different equipments which are geographically distinct and often far away, and whose clocks have neighbouring frequencies (plesiochronous tributaries) or, in the best case, equal (synchronous tributaries), but whose relative phases can certainly be anything and even vary, because the lines have different propagation delays which further depend on temperature.

The multiplexing of the four tributaries requires perfect isochronism between the bits. It is therefore necessary to bring them all to exactly the same rate. This is generally done with the aid of retime buffers, capable of storing an entire frame. The tributary writes the frame in the buffer at its own data rate, while the higher order multiplexer empties the buffer at the new internal rate.

4.4 The T carrier framing and coding formats

4.4.1 The superframe format

The T carrier was designed to carry 24 independent digitised voice channels, with each channel encoded as a 64kbit/s data stream. One of the earliest framing formats was the D4 framing pattern. This T1 frame format evolved principally to carry voice streams, and data to be transmitted over a T carrier system must conform to this format. The frame consists of 193 bits, with the last bit always being a framing bit. The first 192 bits correspond to 24 conversations, or channels, sampled with PCM type methods and generating 8-bit words. The combined signal is word interleaved, providing one frame.

A superframe (ITU-T calls it a multiframe) is a repeating sequence of 12 such frames and thus contains 12 framing/signalling bits. In order to keep track of the frame structure, at least 1 bit in 15 bits of the combined stream (information plus signalling) must be a 1, and at least 3 bits in 24 bits of the stream must be 1s. The bandwidth used on voice frequency (VF) signalling is minimised by putting signalling information only in the LSB in the sixth and twelfth frames.

As illustrated in Figure 4.5, each superframe consists of 12 repeating frames. One frame corresponds to 125 microseconds; one superframe is thus 1.5 milliseconds in duration. Each frame contains one synchronisation bit to allow the receiving equipment to decode, demultiplex, and allocate the incoming bits to the appropriate channels.

From the above follows that each superframe contains a 12-bit word, composed of individual bits coming from each of the 12 frames. The framing bits are called BFf, and the signalling bits BFs. BFfs are the odd numbered framing bits, whilst BFss are the even

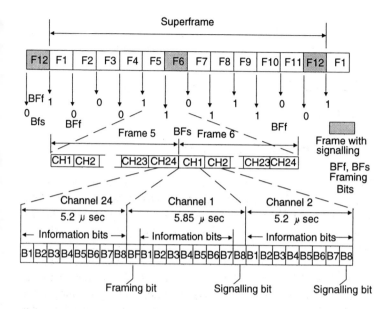

Figure 4.5   D4 framing patterns

numbered bits. The 12-bit word is used for synchronisation and for identifying frames number 6 and 12, which contain channel signalling bits. Each frame thus contains a BFf or BFs framing/signalling bit on the 193rd position. The resulting pattern of the 12-bit word is 100011011100. This entire process repeats every 1.5 milliseconds.

The BFf bit alternates (1, 0, 1, 0, 1, 0, ...) every other frame. The receiving multiplexer can identify this unique sequence in the incoming digital data stream to maintain or re-establish frame boundaries. Frame 6 is identified by the fact that it occurs when the BFs is a 1 preceded by three BFs which were 0s. Frame 12 is identified by the fact that it occurs when the BFs is 0, preceded by three BFs which were 1s. The receiver can easily identify this sequence, in the absence of severe line errors or impairments, and thus identify the desired frames which contain signalling information.

The 193rd bit signalling described above involves the management of the DS1 facility itself. As indicated in Section 4.3.2, each individual voice channel requires its own signalling information (call set-up, call completion, etc.). The signalling information must be transmitted along with the PCM samples. To achieve this the multiplexer will rob, or share the least significant bit, i.e. bit 8 from the user data stream. Consequently, this bit alternatively carries information or signalling data.

For five consecutive frames bit 8 will contain voice bits, and on the sixth, it will contain a signalling bit. This should not be confused with the mechanism described above for the superframe and the 193rd bit, though both mechanism and principle are similar. This bit robbing occurs totally within one voice sample, namely within one octet of bits. The sixth bit is also referred to as the A bit, while the twelfth bit is also called the B bit. These combinations of bits allow the end-user station equipment to carry out its signalling protocol, which involves indicating such states as idle, busy, ringing, no ringing, loop open, etc.

For data applications the A/B signalling has no relevance. However, when transferring data, the impact of losing the LSB every six frames is not acceptable. There are two ways to get around this inherent difficulty. Instead of transmitting data in 8-bit quantities, they are transferred in 7-bit quantities. Therefore, the fact that the LSB is robbed does not affect the overall performance because it is not used. To achieve this, data must be transmitted 7 bits at a time every 125 μsec. In practical terms this involves either using a device that is capable of outputting at 7-bit increments or providing a special timing pulse to enable the data output.

Many communications controller devices can be operated in the latter mode but only a few in the former. As only 7 bits are transmitted every 125 μsec, the data rate is reduced from the 64kbit/s available to 56kbit/s. With this method, 12% of the potential bandwidth is lost (see also Section 4.4.3).

4.4.2 The extended superframe format

The DS1 format described above was developed in the 1960s and is based on technology then available. The basic challenge of DS1 transmission is that of preserving bit and word synchronism. To do so, 24 eight-bit words (corresponding to samples for 24 telephone conversations) are followed by one bit (the 193rd bit) for frame tracking and management. There are 8000 repetitions of this pattern in one second, which results in an additional 8kbit/s bandwidth loss.

The extended superframe format, denoted as ESF, was announced in 1981 and adds two enhancements. First, the number of frames in a superframe is increased to 24, rather than 12. This provides extra signalling bits for telephone signalling, a C bit and a D bit. A 24 frame pattern makes an additional bandwidth available on the framing bit. This is used to send maintenance data between T1 interfaces.

The ESF has 24 frames in its definition of a superframe, but only six bits in its framing pattern. Rather than resynchronising every 1.5 milliseconds as in the regular format, the ESF only needs to resynchronise every 3.0 milliseconds. The 193rd bits are now being looked at as part of one 24-bit word. Substantial progress in VLSI technology has enabled the equipment to keep timing more accurately. This implies that fewer bits are required for this housekeeping function. The result is that 4000bit/s of channel are freed without losing any functionality or any additional bits. This bandwidth can be used for direct communication between intelligent network monitoring devices and centralised network monitoring computers.

Figure 4.6 depicts the ESF format. The successive 193rd bits are shown explicitly. There are 6 bits in the frame synchronisation word, rather than 12. The D bit represents link level data, while the C bit handles error checking and monitoring functions. With the extended superframe format all previous functionality, including the VF signalling rule, remain available. No new bits from the 1.544Mbit/s are taken away from the user.

Since twenty four frames have now to be examined by the equipment to establish synchronisation and extract other channel information, it takes longer to regain lost synchronisation. However, with the newer sophisticated clocking mechanism this should only happen very rarely.

A 4000bit/s data link, also referred to as the Embedded Operations Channel (EOC) or Facilities Data Link (FDL), used for maintenance

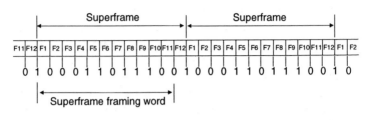

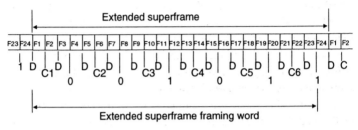

C bits = CRC-6 bits
D bits = Data link bits (maintenance)
001011 = 6 bit framing pattern

Figure 4.6 Extended superframe format

information, supervisory control, and other (future) needs, becomes available.

The ESF format also provides in a 6-bit Cyclic Redundancy Check (CRC). The check character is used for monitoring the transmission quality and overall performance of the DS1 facility. The CRC-6 is generated from the bits of the preceding frame. For the calculation, the framing bits of that frame are considered to equal 1. Since the CRC-6 allows detection of apparent degradation in transmission quality, problems may be fixed before a total failure occurs. The CRC-6 detects about 98.4% of single bit or multiple bit errors. The CRC 6 also provides false frame protection. This follows from the fact that if the wrong synchronisation boundary is selected, the CRC will not calculate correctly.

Under ESF new VF signalling capabilities become available by providing two additional robbed signalling bits, allowing representation of up to 16 states. These bits are known as C and D, which complement the A and B bits of the regular superframe format. Out of these, 16 codes options are available for transparency (no robbed signalling bits), two state signalling (A bit only), four state signalling (A and B bits), or sixteen state signalling (A, B, C, and D bits). In the transparent signalling option, all eight bits within a frame within the extended superframe are given to the user for data. When the A signalling mode is employed, the LSB of every sixth signalling frame is robbed to show the desired bit. When A/B signalling is used, the first and third signalling frames are robbed to carry the A bits, while the second and fourth signalling frames will carry the B bits. If the A/B/C/D signalling is employed, the desired bits are gained by robbing the first, second, third, and fourth frames, respectively.

From the above we can conclude that ESF has economically reallocated the 8kbit/s that have long been used to manage the DS1 facilities. 2000 bits are used for framing (6 bits distributed over 24 framing bits and there are 333 such 24-bit words per second). 2000 bits are used for error detection and performance determination (6 bits distributed over 24 framing bits). Finally, 4000 bits are used for telemetry and facility management and/or reconfiguration (12 bits distributed over 24 framing bits).

4.4.3 Clear channels for data applications

As already mentioned, users of T1 facilities are restricted from accessing the full 64kbit/s bandwidth of the PCM channel on any one circuit of the DS1, whether for voice or data applications, because the eighth bit of each word is unavailable. In voice transmission, the eighth bit has been used for in-band signalling (remember that only the sixth and twelfth frames are really affected). In speech, this generally does not cause any problem, but the same condition cannot be guaranteed for data. The implication has been the deployment of 56kbit/s facilities and a loss of 8kbit/s of usable bandwidth. The maximum data rate in a DDS channel (Bell's Dataphone Digital Service, see Section 4.8) is thus 57kbit/s, with 7 bits per frame used for customer data and the eighth bit reserved for network control. This causes complications, particularly for users interfacing with an international facility, e.g. a 64kbit/s satellite channel.

In many early repeaters that were installed along the transmission cables linking the T1 devices, the synchronisation circuitry was very simple. An oscillator was run at a frequency close to the line pulse stream. The pulses were then used to pull the oscillator on frequency. This method relied upon the availability of pulses in the data stream. Because there could be many repeaters over a T1 span, the cumulative effect of a drift in frequency could cause an unacceptable amount of error in the transmission. Therefore the number of pulses, or the ones density, on the line is important. For voice transmission this is not a problem. However, for data it is a different story. To be able to maintain the ones density, zero code suppression schemes are used. These schemes consist of replacing long sequences of zeros by a special pattern (for details on zero code suppression schemes, refer to Section 4.7).

The assumption on the DS1 stream was that at least one eighth, i.e. 12.5%, of the incoming bits were 1s. If that ratio was not maintained, synchronisation would be lost and an error would result. When the DS1 contains voice channels, work around techniques like changing the eight 0 to a 1 do not create major problems, since the distortion in the voice message is minimal and the human ear would not notice it. In contrast, changing a customer data bit is not acceptable, as that may imply a substantial difference in the received data (e.g. 0001 and 1001 are quite different numbers). Therefore, one bit out of eight in the customer channel is reserved in each frame to inject a 1. This bit is called the network control bit and is not made available to the user. While this resolves the ones density problem, it imposes the speed restriction alluded to above.

The approach embodied in the current generation of equipment, of substituting a 1 in the LSB position of an all zeros word, may be acceptable for voice, because the difference between a 00000000 coded sample and a 00000001 coded sample is imperceptible to the ear. When the bits of a DS1 channel represent directly input digital data, such as DDS or ISDN, however, one is not at freedom to alter the data, and a technical solution must be achieved.

A possible solution has already been introduced. This concept, also referred to as the primary rate interface or PRI, is one of the central physical components of ISDN. Its implementation involves taking all of the signalling information of the individual 23 T1 channels and combining it into a 24th channel.

Often referred to as the "23B + D" format, this framing and signalling scheme allows 23 T1 channels to carry up to 64kbit/s of voice, data, video, or other information, while the D channel transmits all of the control data, including sophisticated signalling techniques described in ITU-T Recommendation I.431/Z.921/Q.931. (Note that he "B" in this designation stands for bearer and that the "D" stands for delta, which is the channel that transmits change information).

4.5 The CEPT PCM-30 framing format

The CEPT PCM-30 is a PCM format used for time division multiplexing of 30 voice or data circuits onto a single twisted pair cable using digital repeaters. As already mentioned, the delay with which CEPT has undertaken the definition of a primary digital PCM system has allowed it to profit from the experience of the American systems.

The basic characteristics of this primary PCM multiplex equipment are described in ITU-T Recommendation G.732. Each voice circuit is sampled at 8 kHz using an 8-bit A-law companding analog to digital converter as specified in ITU-T Recommendation G.711, and multiplexed with 29 other sampled channels plus one alignment and one signalling channel, resulting in 32 multiplexed channels.

The standard CEPT frame contains 32 channels of 8 bits each, or 256 bits. With 8000 samples per seconds, the CEPT data rate becomes 8000 x 256, or 2.048Mbit/s.

4.5.1 Frame composition

As with the DS1 D4 framing format, the CEPT PCM-30 frame structure is subject of ITU-T Recommendation G.704. Each CEPT

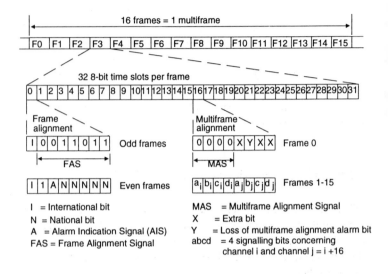

Figure 4.7   CEPT PCM-30 frame and multiframe composition

PCM-30 frame consists of 32 time slots to include 30 voice channels, one alignment signal and one signalling channel. Time slots 1 to 15 and 17 to 31 are assigned to the 30 voice or data channels. This frame composition is depicted in Figure 4.7.

The frame alignment signal FAS (0011011) is transmitted in bit positions 2 to 8 of time slot 0 of every other frame. Bit position 1 carries the International bit, while frames not containing the frame alignment signal, i.e. the odd frames, are used to carry National and International signalling bits and alarm indication for loss of frame alignment. The loss of frame alignment alarm is announced by alarm bit A, which is the third bit of the odd. To avoid imitation of the frame alignment signal, bit 2 of the odd frames is set to 1.

Time slot 16 in each CEPT 30 frame is used to transmit such signalling data as on-hook and off-hook conditions, dialling digits and call progress. Since a common channel is dedicated for the signalling data of all voice circuits, this method of signalling is referred to as common channel signalling. The signalling consists of four bits per channel, grouped in the two halves of time slot 16. It therefore requires 15 frames to carry the information of the 30 channels. Completed by a 16th frame (frame 0) they constitute a multiframe.

Time slot 16 of this 16th frame contains a multiframe alignment pattern MAS which allows unambiguous numbering of the frames within the multiframe. Each channel thus arranges the signalling of 4 bits every 16 frames, i.e. every 2 milliseconds.

4.5.2 CRC-4 cyclic redundancy check

Where there is a need to provide additional protection against simulation of the frame alignment signal, and/or where there is a need for an enhanced error monitoring capability, then bit 1 of the frame, i.e. the International bit, is used for the 4-bit cyclic redundancy check described below.

Each CRC-4 multiframe, which is composed of 16 frames numbered from 0 to 15, is divided into two 8 frame sub-multiframes or SMF, designated SMF I and SMF II which indicates their respective order of occurrence within the CRC-4 multiframe structure.

The SMF is the Cyclic Redundancy Check-4 block size, i.e. 2048 bits.

In the frames containing the frame alignment signal, bit 1 is used to transmit the CRC-4 bits. There are four CRC-4 bits, referred to as C1, C2, C3, and C4 in each SMF. In the frames not containing the frame alignment signal, bit 1 is used to to transmit the CRC-4 multiframe alignment signal and two CRC-4 error indication bits (E). The CRC-4 multiframe alignment signal is a 7-bit sequence having the form 0011011.

The E-bits are used to indicate received errored sub-multiframes by changing the binary state of one E-bit from 1 to 0 for each errored sub-multiframe. Recommendation G.704 requires the delay between the detection of an errored sub-multiframe, and the setting of the E-bit that indicates the error state, to be less than 1 second.

Table 4.3 shows the allocation of bits 1 to 8 of the frames for a complete CRC-4 multiframe.

Table 4.3 CRC-4 multiframe structure. (E = CRC-4 error indication bits; S_4 to S_8 = Spare bits; C_1 to C_4 = CRC-4 bits; A = alarm indication signal (AIS)

	Sub-multi-frame (SMF)	Frame number	Bits							
			1	2	3	4	5	6	7	8
M		0	C_1	0	0	1	1	0	1	1
U		1	0	1	A	S_4	S_5	S_6	S_7	S_8
L		2	C_2	0	0	1	1	0	1	1
T	SMF I	3	0	1	A	S_4	S_5	S_6	S_7	S_8
I		4	C_3	0	0	1	1	0	1	1
		5	1	1	A	S_4	S_5	S_6	S_7	S_8
		6	C_4	0	0	1	1	0	1	1
		7	0	1	A	S_4	S_5	S_6	S_7	S_8
F		8	C_1	0	0	1	1	0	1	1
R		9	1	1	A	S_4	S_5	S_6	S_7	S_8
A		10	C_2	0	0	1	1	0	1	1
M	SMF II	11	1	1	A	S_4	S_5	S_6	S_7	S_8
E		12	C_3	0	0	1	1	0	1	1
		13	E	1	A	S_4	S_5	S_6	S_7	S_8
		14	C_4	0	0	1	1	0	1	1
		15	E	1	A	S_4	S_5	S_6	S_7	S_8

4.6 T1 and PCM-30 alarms and error conditions

4.6.1 Principal alarms

The principal alarms defined by both T1 and PCM-30 are the red alarm, generated by the receiving equipment to indicate that it has lost frame alignment, and the yellow alarm, which is returned to the transmitting terminal to indicate that the receiving terminal has lost frame alignment. Normally, the terminal will use the receiver's red alarm to request that a yellow alarm be transmitted. The name of these alarms simply comes from the colour of the lights on the original equipment.

Loss of frame alignment is detected simply by monitoring the frame alignment signal. PCM-30 differentiates between loss of frame alignment, being a failure to synchronise on the frame alignment signal FAS (0011011), and loss of multiframe alignment, which is caused by a failure to synchronise on the multiframe alignment signal (0000) contained in bits 1 to 4 of time slot 16 of frame 0. The former is the red alarm, the latter is the multiframe red alarm. Table 4.4 illustrates how these alarm conditions are transmitted. T1 multiplexers can also signal a blue alarm. This is a continuous ones pattern across all 24 channels (the F-bits, however, remain unchanged) to indicate an upstream failure.

4.6.2 Error conditions

The most basic impairment that T1 or E1 equipment can suffer from are bipolar violations (BPVs). A bipolar violation is a violation of bipolar coding in which two pulses occur consecutively with the same polarity. As will be discussed later in this chapter, T1 and E1 signals are encoded with a system that inverts the polarity of alternate one bits so that two pulses of the same polarity will not occur in a row. On metallic circuits, a bipolar violation will occur if a zero is changed to a one, or if a one is changed to a zero. Since bipolar violations occur

table 4.4   Transmission of alarm conditions

Alarm	Multiplex mode	Format
Transmitted Yellow Alarm	T1 D4	Bit 2 = 0 in all channels and BFs = 1 in frame 12
	T1 ESF	Repeated pattern of 8 zeroes 8 ones on data link
	PCM-30	Bit 3 in TSO of non-frame alignment frame is set
Yellow Alarm at receiver	T1 D4	Bit 2 = 0 for 255 consecutive channels and BFs = 1 in channel 12
	T1 ESF	16 patterns of 8 zeroes 8 ones on data link
	PCM-30	Bit 3 in TSO on non-frame alignment frame is set

one for one with bit errors, the BPV rate (the ratio of BPVs to correct bits) corresponds to the bit error rate.

CRC errors are a second possible T1/E1 error condition. The cyclic redundancy check error measurement is an alternative to frame error measurement, and is available to T1 circuits that employ the Extended Superframe (ESF) format and to E1 circuits that employ the CRC-4 multiframing.

CRC-n (n = 4 with PCM-30, n = 6 with T1 ESF) is an error checking method that uses an n-bit code to represent an entire multi-frame of data bits. The n-bit code is arrived at by applying a complex mathematical function to each group of 24 (ESF) or 8 (PCM-30) frames of data. The result of this calculation, the n-bit code, is then transmitted in the CRC framing bit positions of the following frame.

At the other end of the circuit, the same mathematical function is performed on the same first group of frames. This newly calculated n-bit code is compared with the code that was calculated by the transmitting equipment. Any discrepancies between the two codes are counted as CRC errors. CRC errors are by far the best in service performance measurements because the CRC scheme allows detection of errors on all of the data bits within each group of frames, with an accuracy of about 98 percent.

Slips are yet another impairment. Slips are the most common impairments caused by frequency deviations and timing problems. A slip is the insertion or deletion of data bits into or from the data stream. It is the direct result of equipment buffer overflow or underflow, resulting from improperly timed network equipment. Digital equipment uses input buffers of finite length to accommodate the momentary frequency fluctuations that can occur between the receiver and transmitter of a digital network node. These frequency differences become persistent when connecting to (or selecting) the wrong clocking source, causing buffers to overflow or underflow and then reset themselves. It is the buffer resetting that results in the addition or deletion of data bits from the bit stream.

Based on the source of the slips and their effects on the network, all slips can be placed in one of two categories, controlled slips or uncontrolled slips. Controlled slips are bit additions or deletions that do not disrupt frame synchronisation. Uncontrolled slips are bit additions or deletions that cause both framing and data to be displaced. This framing and data displacement results in a loss of frame synchronisation, effectively taking the circuit down momentarily.

Jitter, the cyclic offset of bits from their expected positions in time, is one of the most ominous of all T1/E1 impairments. Jitter can be intermittent and data dependent, which makes it difficult to isolate. Jitter occurs little by little, cumulatively over many bits, and can ultimately cause the missampling of the pulses resulting in bipolar violations and bit error conditions.

The most common cause of jitter is the network equipment itself. Jitter is inherent to the clock recovery timing used in transmission, and is typically added to the pulses at every regeneration point within the network. As long as each network component adds only a very small amount of jitter, the circuit will be unaffected. Problems arise when a failed or failing network component adds significant amounts of jitter. Less typically, jitter can come from crosstalk, electrical noise, and other types of interference.

Wander is an impairment very similar to jitter, and is defined as jitter occurring at a frequency of less than 10Hz. Wander, like jitter, is a back and forth (cyclic) displacement of bits from their expected positions. But, because wander occurs at such low frequencies, its cause within the network and effects on the network are very different from those of jitter. Wander is most often caused by instabilities in a master timing source, or by nocturnal cooling (cooling as the sun goes down). The end result is usually slips.

4.7 Coding schemes

Another aspect of digital transmission systems is the algorithm used for data transmission on the line. There are various ways in which the data can be represented as signals on the line. One of the important factors of a transmission system is the number of discrete levels a signal can have. For example, TTL has two levels to encode the binary data. One voltage level indicates a logic 1, and the other voltage level a logic 0. A single switching point, or decision level, is used to set the threshold. By comparing the coded signal to the threshold level, the binary information can be decoded. If the signal level is above the threshold, then a logic 1 is detected; conversely, if it is lower, a logic 0 is detected. Because the signal has two levels, it is known as a unipolar code. Unipolar coding is a very good system for TTL systems but is not suitable for transmission systems using copper cables over long distances.

The reason why lies in the construction and characteristics of such cables. A transmission cable is made of a group of wires, having two electrical properties. One is the d.c. resistance and the second is the self inductance value of the wire. When a group of wires is placed in a casing to form the cable, a third electrical property, capacitance between the wires, is brought to bear. The transmission cable now behaves like a low pass filter.

If a unipolar code is transmitted down the line, each high level signal will inject energy into the capacitor/inductor of the cable. Conversely, a low level signal will discharge the line. If the number of highs and lows is matched, then the net energy level on the line will be zero. But if the number is not matched (which is normally the case), the transmission line will have periods during which energy is stored. This stored energy will result in a d.c. offset being superimposed on the transmitted signal. The d.c. offset will interfere with the decoding of the received pulse by reducing the difference between the high and low signal levels and the threshold level. For instance, if the threshold level is 1.3V and a low level signal is 0.8V, then a DC offset of greater than 0.5V will stop the decoding of a low level signal and affect the overall performance of the transmission system.

To overcome this, transmission codes with three levels, positive high, zero, and negative low, have been developed. The highs and lows are used to represent the same logic level. The main difference between this system and the unipolar system is that the highs and lows are alternated for the same logic level. For example, in the system of coding known as alternate mark inversion (AMI), a high or a low level represents the logic level 1. A zero level represents the logic level 0. To transmit the sequence 10101 the following pattern would be output: high/zero/low/zero/high. This type of coding is known as pseudoternary coding. It is called "pseudo" because the zero voltage level is not really classified as a discrete level. The receiver needs two decision levels to decode the incoming data and also has to keep track of the level of the last logic 1 transmitted so that the correct level of the next logic 1 is sent as the code alternates. In the case in which the wrong level is detected, i.e. a high followed by a high, a code violation is recorded. The advantage of this type of coding is that the d.c. level, or balance, is maintained by the transmission of alternate high and low pulses. The disadvantage is the extra circuitry required.

Many receivers extract clocking information from incoming data by detecting the occurrences of the incoming signal data crossing a decision threshold. Each time the signal crosses this point, a PLL can lock its output to it. To perform this task effectively there have to be sufficient crossings over a period of time. If the number of crossings is reduced, the PLL can drift from the frequency of the incoming signal. This will cause the incoming signal to be sampled incorrectly and result in data errors. With a ternary code, the logic levels encoded into alternate high and low output signals provide the clocking information. For example, in AMI coding, each time a logic 1 is transmitted the receiver can lock the PLL onto the incoming signal.

Unfortunately, long series of zeros, which are impossible to prevent, may deprive the PLL of all synchronisation information. To avoid this, a group of several consecutive zeros is replaced by a group containing a factitious 1, signalled as such to the receiver by a polarity in violation of the law of alternation of the AMI mode. This principle is systematically applied to all primary and second order digital transmission systems.

The European digital system makes use of a pseudo-ternary mode called a high density bipolar 3-zero maximum code HDB3, which avoids the appearance of more than three consecutive zero symbols, as illustrated in Figure 4.8. It consists of replacing groups of four binary zeros by groups of four ternary symbols of which the the last is non-zero and transmitted with the same polarity as the last non-zero symbol, i.e. in violation of the alternation law of the AMI code. This allows the easy identification of such group at reception and its interpretation as four binary zeros. Furthermore, the first of the four ternary symbols is chosen to be positive, zero, or negative in such way as to maintain or reset the d.c component to a zero value.

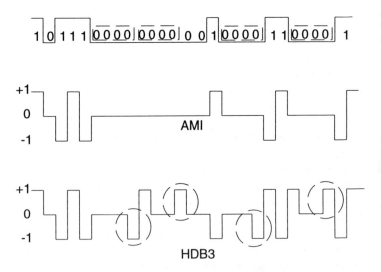

Figure 4.8 AMI and HDB3 pseudo-ternary codes

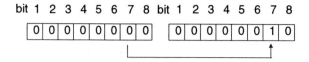

Figure 4.9 Typical B7 zero code suppression example

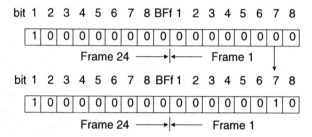

Figure 4.10 B7ZS worst case scenario

The American T1 system generally uses a technique called B7 zero code suppression (B7ZS). To keep telephone company T1 line repeaters and channel service units (CSU) in synchronisation, the digital bit stream permits a maximum of 15 consecutive zeros. As already mentioned, this is known as the Telco's ones density requirement (also refer to Section 4.4.3). To obtain compliance with the ones density requirement, communications carriers use the B7ZS technique.

As an example of B7 zero code suppression, consider the DS0 time slot illustrated in Figure 4.9. If all 8 the bits are zeros, B7 zero code suppression will substitute a 1-bit in position 7. The adapted time slot is also depicted in Figure 4.9.

Figure 4.10 shows the worst case scenario, where channel 24 is followed by a 0 frame bit, and all bits in channel 1 are zeros, resulting in a total of 16 consecutive zeros. The illustration shows that in this case B7 zero code suppression reduces the number of consecutive zeros to 14.

As already mentioned in Section 4.4.3, this coding technique followed from the assumption that, in order to guarantee synchronisation, at least one eighth, i.e. 12.5%, of the incoming bits were 1s. However, if a data channel contains all 0s, the data will be corrupted due to B7 zero code suppression. As a result, a data channel is normally restricted to seven usable bits, with one bit set to a 1. This prevents the user data from being corrupted but limits the actual bandwidth to only 56kbit/s. When one bit is set to a 1 on a DS0 channel, the channel is said to be a non-clear channel. The 56 kbit/s on a non-clear channel is also known as a DS-A channel.

Clear channel capability can be obtained by using bipolar transmission with bipolar 8 zero substitution (B8ZS) in the T1 bit stream (Figure 4.11). With B8ZS no more than eight logic 0s can be transmitted sequentially. Therefore, the data stream to be transmitted is examined to determine if a long sequence of logic 0s is about to be transmitted. If such a sequence is detected, each eight consecutive 0s in a byte are replaced by a special pattern. If the pulse preceding the all zero byte is positive, the inserted B8ZS code is $000 + - 0 - +$. If the pulse preceding the all zero byte is negative, the inserted B8ZS code is $000 - + 0 + -$. Both examples result in bipolar violations occurring in the fourth and seventh bit positions. This special pattern is unique because of the embedded code violations. When detected at the receiving side it is removed and not seen as a code violation.

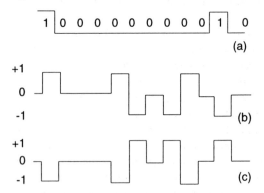

Figure 4.11   Bipolar 8 zero substitution coding

If a string of nine logic 0s were to be transmitted, then the first eight logic 0s would be replaced by the special pattern while the ninth 0 would be transmitted normally. At the receive end, the special pattern, which is expected by the receiver, is decoded back to the eight zeros removed. The data output by the receiver is still the nine logic 0s that were to be transferred. Two things can be seen from the example. The addition of the extra, special pattern gave the receiver the clocking data it required to allow the PLL to maintain synchronisation. Secondly, to transmit the sequence of nine binary digits still requires only eight pulses on the line. Although the receiver will be less likely to lose synchronisation, more complicated transmitters and receivers have to be used.

4.8 Digital network services

This section provides a brief description of three digital network facilities that are based on T1 and PCM-30 time division multiplexers. Bell's Dataphone Digital Service (DDS) and AT&T's Accunet T1.5 are typical examples of T carrier multiplexer networks, while BT's KiloStream and MegaStream networks are based on the European standards.

4.8.1 Bell's DDS subrate facility

Bell's Dataphone Digital Service (DDS) was approved by the U.S. Federal Communications Commission in December 1974. Currently, there are over 100 cities connected to the DDS network in the United States as well as international connections to other digital networks.

DDS is an all synchronous facility. Currently supported transmission rates and services include 2400, 4800, 9600 and 56000bit/s leased circuits. Recently a 56kbit/s switched service has been added. For transmission at different data rates, specialised equipment to include multiplexers and/or converters must be employed.

The carrier structure of the DDS network is illustrated in Figure 4.12. DDS facilities are routed from a subscriber's location to an Office Channel Unit (OCU) located in the carrier's serving central office.

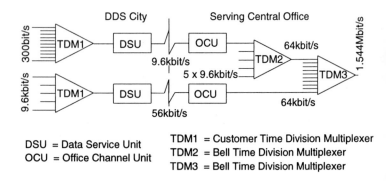

Figure 4.12 DSS carrier structure and multiplexing arrangement

The signalling structure used on DDS facilities is a modified bipolar signalling code. The modification to bipolar return to zero signalling results in the insertion of zero suppression codes to maintain synchronisation whenever a string of six or more consecutive zeros is encountered.

It should be clear by now that precise synchronisation is the key to the success of an all digital network. Timing ensures that data bits are generated at precise intervals, interleaved in time and read out at the receiving end at the same interval to prevent the loss or garbling of data. To accomplish the necessary clock synchronisation on the AT&T digital network, a master clock is used to supply a hierarchy of timing in the network. Should a link to the master clock fail, the nodal timing supplies can operate independently for up to 2 weeks without excessive slippage during outages. In Figure 4.13 the hierarchy of timing supplies as linked to AT&T's master reference clock is illustrated. As shown, the subsystem is a tree-like network containing no closed loops.

Within a DDS office, a composite timing signal is distributed over balanced pairs to each bay of equipment. This timing signal is a bipolar 64 kHz waveform having a 5/8 duty cycle. Each eighth pulse violates the bipolar rule. Hence, the basic waveform provides the bit

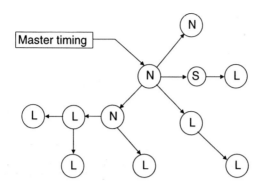

N = Nodal Timing Supply
L = Local Timing Supply
S = Secondary Timing Supply

Figure 4.13 DDS network timing

clock information while the bipolar violation provides the byte clock information.

Connection to the DDS service originally required a channel service unit (CSU) and a data service unit (DSU) to terminate a DDS line. The DSU converts the signal from the terminal equipment into the bipolar format used with DDS and T1 facilities. The CSU performs line conditioning functions to include equalisation and signal reshaping as well as line loopback testing. Prior to deregulation, the CSU was provided by the communications carrier, while the DSU could be obtained from third party sources. After deregulation, several third party vendors started manufacturing devices that combined the functions of the DSU and the CSU.

4.8.2 BT's KiloStream

KiloStream is a point to point leased line digital service that was first offered commercially in January 1983. Like DDS, KiloStream is an all synchronous facility. BT provides a Network Terminating Unit (NTU) which is similar to a DSU/CSU to terminate the subscriber's

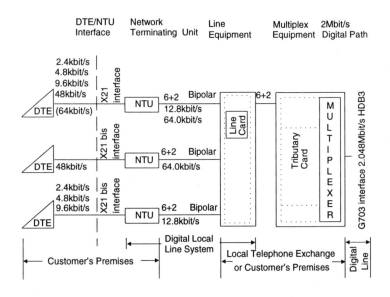

Figure 4.14 Basic KiloStream structure

line. The NTU encodes customer data for transmission via a digital local line to the local exchange where it is fed into a CEPT 2.048 Mbit/s multiplexer, as illustrated in Figure 4.14.

The customer data rate can be 2400, 4800, 9600, 48000 or 64000bit/s. The DTE/NTU interface depends on the data rate. Speeds of 2400, 4800, and 9600bit/s can be serviced through a ITU-T X.21 or ITU-T X.21 bis/V.24 interface. 48kbit/s can be provided with a ITU-T X.21 or ITU-T X.21 bis/V.35 interface. A 64kbit/s channel is only available with an X.21 interface. At all data rates, except at 64kbit/s, the NTU encodes the user data into a 6+2 envelope structure as described in ITU-T Recommendation X.50. Hence, the line data rate is higher than the customer data rate. For DTE data rates of 2400, 4800, and 9600bit/s the line data rate equals 12.8kbit/s, while a 48kbit/s user channel is transported at a 64kbit/s line rate.

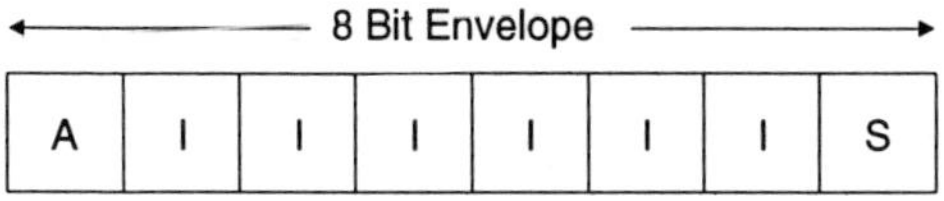

A = Alignment bit which alternates between 1 and 0 in successive
    envelopes to indicate the start and stop of each 8-bit envelope

S = Status bit which is set or reset by the control circuit and checked
    by the indicator circuit

I = Information bits

Figure 4.15 KiloStream 8-bit envelope encoding

Customer data is framed into a 6+2 format to provide the signalling and control information required by the network for maintenance assistance. This is known as envelope encoding and is illustrated in Figure 4.15. The rate of 8000 octets/s, imposed by the 8kHz sampling frequency of PCM systems, allows a total net rate of 48kbit/s per digital channel. One such channel could, for example, transport 5 sub-channels at 9600 bit/s, 10 sub-channels at 4800bit/s, or 20 sub-channels at 2400bit/s. ITU-T Recommendation X.58 optimises this structure by minimising the signalling overhead and is able to transport up to 6 sub-channels at 9600bit/s.

The NTU provides a ITU-T interface for customer data at 2.4, 4.8, 9.6 or 48kbit/s to include performing data control and supervision, which is known as structured data. At 64kbit/s, the NTU provides a ITU-T interface for customer data without performing data control and supervision, which is known as unstructured data.

The NTU controls the interface via ITU-T Recommendation X.21, which is the standard interface for synchronous operation on public data networks. An optional V.24 interface is available at 2400, 4800 and 9600bit/s, while an optional ITU-T V.35 interface can be obtained at 48kbit/s. With the X.21 interface, the control circuit (C) indicates the status of the transmitted information, data or signalling, while the indication circuit (I) signals the status of information received from the line. The control and indication circuits control or check the status bit of the 8-bit envelope used to frame six information bits.

4.8.3 AT&T's Accunet and BT's MegaStream

Both Accunet T 1.5 and MegaStream are carrier transmission facilities for high speed data and high volume speech communications. The Accunet T1.5 facility operates at 1.544Mbit/s. The MegaStream facility operates at 2.048Mbit/s. Both facilities offer full or fractional T1/E1.

Data sources that can be effectively serviced by these transmission facilities include digital PABXs, analog PABXs with voice digitisers, clustered terminals via multiplexer input, analog terminations, and video applications.

4.9 References

Bell (1982). *Bell System Technical Reference*, PUB 43801, November.

Bocker, P. (1988) *The Integrated Services Network*, Springer Verlag (West Germany).

Bylanski, P. and Ingram, D.G.W. (1976) *Digital Transmission Systems*, Peter Peregrinus, Stevenage.

CCITT (1989) *Blue Books, Volume III – Fascicle III.4, General Aspects of Digital Transmission Systems – Terminal Equipments, Recommendations G.700 – G.795*, CCITT, Geneva.

Lucky, R.W., Salz, J., Weldon Jr, E.J. (1968) *Principles of Data Communication*, McGraw-Hill, New York.

Matick, R.E. (1969) *Transmission Lines for Digital and Communication Networks*, McGraw-Hill, New York.

5. Digital transmission

5.1 Design principles

5.1.1 System requirements

A simplified block diagram of a digital communication system is shown in Figure 5.1. Information from a data source is processed so that it can be reliably transmitted via a communication channel to a distant data terminal. The user will measure reliability in terms of the difference between the transmitted data and the received data. The system designer's task is to ensure that this difference is always acceptably low.

The processing tasks in the transmitting terminal fall into two categories:

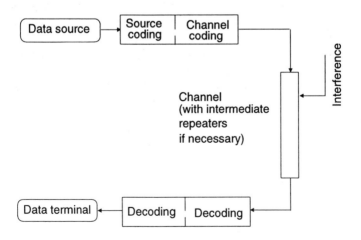

Figure 5.1 A digital communication system

1. Source coding is concerned with the structure and statistics of the data source. It will involve the conversion of the data into a more suitable form for subsequent processing. Word alignment and frame alignment (when multiplexing several data sources together) can also be regarded as part of the source coder's task.

2. Channel coding is concerned with the characteristics of the transmission channel, it must ensure that the processed data is compatible with the requirements of the channel. For example, it may be necessary to add redundancy to the transmitted data symbols to combat the effects of transmission errors by providing some error detection and/or correction capability. Symbol timing information may also need to be added. In addition, coding must ensure that the frequency spectrum of the processed data is compatible with that of the channel, thus line coding for low-pass channels and modulation (carrier keying) for band-pass channels, are also tasks for the channel coder.

In practice it is not always possible to make a clear distinction between source and channel coding functions. Indeed, improved efficiency can sometimes be obtained by combining some tasks (Hamming, 1980).

We now concentrate on the fundamental aspects of digital transmission with emphasis on transmission over low-pass channels. These same principles apply to band-pass channels which are treated elsewhere.

5.1.2 The baseband waveform

Figure 5.2(a) shows a binary data signal, assumed to be in a suitable form for transmission over a low-pass channel, this is often referred to as the baseband waveform. After transmission over the restricted bandwidth of a channel (and if necessary, some equalisation to partially compensate for the characteristics of the channel) the signal may look like Figure 5.2(b).

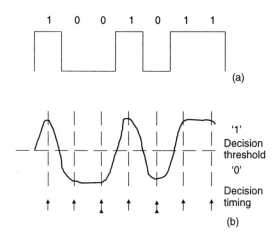

Figure 5.2   Baseband waveforms: (a) transmitted signal; (b) bandlimited received signal

The task of the receiver is to establish the transmitted sequence. For a binary sequence, a single decision threshold is required. Each amplitude decision will have to be made in the presence of any distortion and interference superimposed upon the data sequence during transmission. The decision making process will be more reliable if it is timed by a clock which distinguishes when each received symbol has reached an optimum value. The accuracy with which the decision threshold and sample times must be placed will depend upon the severity of distortion suffered during transmission. Wideband channels lead to less critical amplitude and time placement requirements, but they may admit more noise into the receiver. A compromise must therefore be struck.

The tasks of equalisation, clock extraction and decision making are shown in Figure 5.3. Also shown is the output stage which, under the control of the extracted clock, produces a retimed baseband waveform. Most digital transmission systems use such a self-timed fully regenerative arrangement at the distant terminal. A similar configuration is common at intermediate repeaters (if used), although untimed amplifiers are being considered for some optical systems.

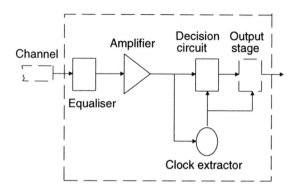

Figure 5.3   A baseband regenerator

5.1.3 Noise and the decision process

The effect of noise on the decision process is treated in many textbooks (Carlson, 1986; Sklar, 1988; Lee, 1988). For zero-mean Gaussian noise, the probability of making an erroneous decision is given by Equation 5.1, where a is the voltage difference between the received signal and its associated decision threshold at the decision time and σ is the r.m.s. noise voltage.

$$P_e = \frac{1}{\sigma\sqrt{2\pi}} \int_a^\infty e^{-x^2/2\sigma^2} \, dx \tag{5.1}$$

This is known as the Gaussian tail area formula and is often shortened to Equation 5.2.

$$P_e = T\left(\frac{a}{\sigma} \right) \tag{5.2}$$

Unfortunately, there is no analytic solution to this equation and so tables (Carlson, 1986; Sklar, 1988), bounds or approximation formulae (Beaulieu, 1989) have to be used.

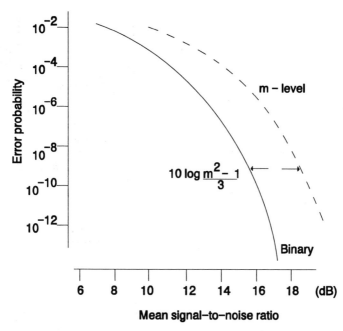

Figure 5.4 Error probability due to Gaussian noise

The solid line of Figure 5.4 shows the error probability (P_e) plotted as a function of the signal-to-noise ratio $20 \log_{10} \frac{a}{\sigma}$ at the input to the decision circuit. This line is equivalent to the case of binary transmission. For an m-level transmitted signal (requiring m − 1 decision thresholds at the receiver), it can be shown (Bennett and Davey, 1965) that for a given symbol error probability an improvement in signal-to-noise ratio of approximately (m^2 − 1)/3 is required relative to the binary case. This analysis assumes a constant mean signal power and that all transmit levels are equiprobable.

The steepness of the curve in Figure 5.4 means that small changes in signal-to-noise ratio have a significant effect on error probability. Thus digital systems need an adequate design margin to ensure both short and long term reliability. We also see that multi-level systems require much better signal-to-noise ratios. This may be partially compensated by the narrower bandwidth requirements made possible by a trade-off between the number of transmit levels and the symbol rate. However for baseband transmission the balance is usually in favour of binary or 3-level transmission at the most.

5.1.4 Waveform shaping and bandwidth requirements

In order to use the transmission medium efficiently and to reduce received noise, the channel bandwidth should be minimised. However, this conflicts with the requirement to restrict the spreading of the pulse waveform at the receiver.

For reliable decision making, a transmitted symbol must be shaped by the channel and its associated equaliser so that it does not interfere with other received symbols. Figure 5.5 shows a common pulse shaping objective. Symbols are transmitted with time spacing T and shaped before presentation to the decision circuit to yield a peak (at the time a decision is made) with zero amplitude at all other decision times. Such a signal is said to exhibit zero intersymbol interference (i.s.i.) with respect to neighbouring received signals.

Nyquist (Nyquist, 1928) showed that the minimum transmission bandwidth (channel plus equaliser) which meets this condition is one which passes all frequencies up to 1/2T and stops all others, as

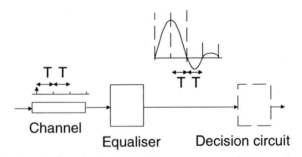

Figure 5.5 Waveform shaping for zero intersymbol interference (basic requirements)

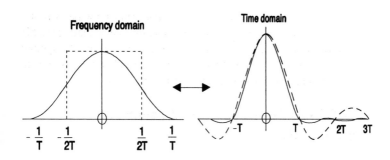

Figure 5.6 Waveform shaping for zero intersymbol interference (Nyquist shown in dotted line and full raised cosine in solid line)

illustrated by the dashed lines of the frequency and time domain pair of Figure 5.6. This result is only of theoretical interest because a transmission characteristic with an infinitely sharp cut-off cannot be realised in practice.

Furthermore, small errors in decision timing can lead to large amounts of i.s.i. Both problems can be overcome, but only at the expense of extra bandwidth. A popular solution, which preserves the zero crossing points, is to provide an equaliser such that the overall transmission characteristic has a raised cosine shape in the frequency domain. A full (100%) raised cosine is shown as the solid lines on Figure 5.6. This is seen to have a compact time response in that it always remains within the magnitude of the Nyquist pulse envelope. Thus, at the cost of more bandwidth, tolerance to (small) timing errors is good.

The full raised cosine has the added feature that the waveform at time T/2 is half the peak amplitude. This leads to good eye shape (see next section), and also has application to some partial response systems (Bennett and Davey, 1965). Other pulse shaping strategies which aim to give good timing tolerance are to be found in Franks (1968).

It should be noted that the characteristics of Figure 5.6 imply linear phase properties. Deviations from this condition can lead to severe waveform distortion (Sunde, 1961). Equalisers must therefore take account of both phase and amplitude. Practical solutions, which can only approximate to the theoretical requirements, aim to give acceptably low (but usually not zero) i.s.i.

5.1.5 The eye diagram

Degradations in digital systems fall into two categories:

1.   Deterministic degradations such as errors in equalising, offsets in decision timing, gain errors and possibly transmission echoes.
2.   Stochastic degradations such as noise, interference, crosstalk and timing jitter.

Some degradations, it could be argued, should appear in both categories.

The deterministic degradations are conveniently assessed by means of an eye diagram. This is obtained, on an oscilloscope for example, by writing all possible received sequences on top of each other whilst triggering the oscilloscope timebase from the data clock as illustrated in Figure 5.7(a) and (b).

The eye diagram provides a great deal of information about the performance of a digital system. In the absence of noise, the width of the eye opening gives the time interval over which the received signal can be sampled without error from i.s.i. The sensitivity of the system to decision timing errors is determined by the rate of closure of the eye as the sampling time is varied. The height of the eye, at a specified decision time, determines the margin over noise.

Given distribution probabilities for the noise and timing jitter, it is then possible to determine the optimum sample time and decision threshold voltage. Figure 5.7(c) gives a diagrammatic view of these points. The diagram serves to illustrate how the worst case signal margin can be very much less than the ideal design value. Noting the steepness of the error probability curve (Figure 5.4), it can be seen how worst case conditions can dominate performance. It follows that reliable estimates for the worst case signal margin are crucial if the performance of a system is to be predicted accurately.

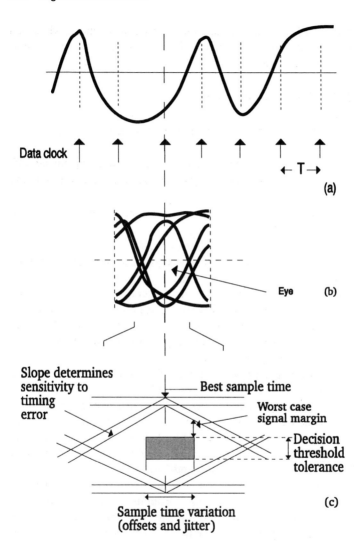

Figure 5.7 Construction of a binary eye: (a) equalised binary signal; (b) eye diagram; (c) interpretation of the eye

5.2 Transmission error performance

5.2.1 The need for performance measures

Transmission errors are usually the most significant impairment to be found in a digital communication system. They are also difficult to predict and therefore difficult to quantify. Consequently, the study of their occurrence and the means of specifying their characteristics has been and continues to be the subject of considerable activity; see for example Yamamoto and Wright (1989) and references therein.

The principal causes of transmission errors in a communication system are:

1. Noise, in particular thermal noise for metallic cable and radio systems, and quantum noise for optical systems.
2. Impulsive interference, often from nearby electro-mechanical equipment.
3. Cross-talk and intermodulation, from other interfering transmission systems.
4. Echoes and signal fading, arising from mismatches in transmission channels and from multi-path effects in radio systems.
5. Terminal equipment limitations and misettings, such as errors in equalisation and decision timing.
6. Network problems, including frame synchronisation errors and lost bits.

Various methods have been used to model these causes of degradation and interference (Kanal and Sastry, 1978; Knowles and Drukarev, 1988). Suffice it to note here that at the receiving terminal or regenerator in a digital link, the probability of making an erroneous decision (the error probability) is critically dependent upon the received signal-to-interference ratio. A small change in this ratio will result in a large change in error probability. It is thus vital to establish error performance measures so that both users and transmission system providers can anticipate an end-to-end performance which is compatible with the requirements of the digital information to be carried. Such performance standards are now emerging, from the

ITU-T (formerly CCITT) for example. Recommendations for error performance are often based on three performance measures, bit-error ratio, available time and error-free seconds, as described in the following sections.

5.2.2 Bit-Error Ratio (BER)

The bit-error ratio (CCITT, 1988a) sometimes referred to as the bit-error rate, is defined as the number of bits received in error divided by the total number of bits transmitted in a specified time interval. Within the specified interval, it is numerically equal to the bit-error probability.

Studies have been conducted to quantify the BER requirements for various digital services (Yamamoto and Wright, 1989). Some typical objectives, expressed in terms of a long term mean BER, are given in Table 5.1.

However, the long term mean BER will usually be an insufficient performance measure by itself. For many digital services the degree of burstiness of the errors, that is their distribution, will also be important. Figure 5.8 shows a typical result where, for a given long term mean BER, the BER over shorter measurement intervals is seen to exhibit considerable variation with time. During some measure-

Table 5.1 Some typical BER objectives

Digital service	Transmission rates (approx.)	Long term mean BER
Voice: Log-law PCM	64kbits/s	2×10^{-5}
Voice: ADPCM	32kbits/s	10^{-4}
Video: Linear PCM	60Mbits/s	2×10^{-7}
Video: Inter-frame coding	2Mbits/s	10^{-10} (approx.)
Data	16kbits/s to 600Mbits/s	10^{-7} (approx.)

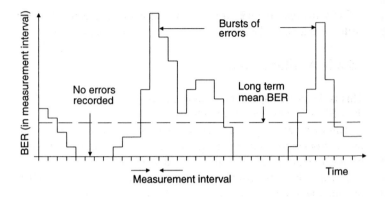

Figure 5.8 Variation of BER with time in a burst error environment

ment intervals no errors are recorded whilst at the other extreme bursts of interference can cause bursts of errors and thus a high short term BER.

Within the illustrative limits quoted in Table 5.1 speech, video and data services will have quite different long and short term BER requirements. For speech, a listener will usually be tolerant to the low level of background noise caused by randomly distributed errors (Takahashi, 1988). Burst errors usually cause more objectionable audible clicks but are still tolerable if they do not occur too often. Errors in a digital video signal will impair quality (CCIR, 1986), and again a viewer will usually be happier with the effects of random rather than burst errors. With video however, additional care is needed as excessive bursts of errors can result in a sudden and catastrophic loss of picture synchronisation. A similar sharp bound on acceptability may also occur with data if error control coding is used to protect the data from transmission errors. Occasional errors in an encoded block or even bursts of limited duration may be acceptable (in a retransmission scheme for example). However, a point will come when the error control coding system can no longer cope and then significant numbers of decoded data blocks may contain errors, now the proportion of error free blocks becomes a useful measure of performance.

Thus, in order to quantify the error distribution characteristics, the BER measure must be refined by the addition of other measures.

5.2.3 Available time

This is the percentage of time, in a specified time interval, in which the BER is less than a specified threshold.

This measure is especially useful when blocks of data are protected by an error control coding scheme. The coding algorithm will usually be able to handle a given number of errors in a block.

Blocks with more errors will be decoded erroneously. If the BER measurement interval corresponds to the error control block length then, with the appropriate BER threshold, the available time will correspond to the percentage of error free blocks after decoding the received data.

5.2.4 Error-Free Seconds (EFS)

This is the proportion of one-second intervals, in a specified time interval, in which the transmitted data is delivered error free. Again it is usually expressed as a percentage.

For modern high bit rate transmission systems a second may be too long (it will contain many bits) and so an error-free millisecond, and error-free microsecond or more specifically, an error free data block (as mentioned above) may become more appropriate measures.

5.2.5 Performance standards — an example

The parameters BER, available time and EFS and adaptations of them are used in the emerging error performance standards for digital links (McLintock and Kearsey, 1984). The important ITU-T Recommendation G.821 (CCITT, 1988a) serves as an example. It specifies error performance objectives for long distance 64kbit/s data connections (Table 5.2).

These are long term objectives, usually measured over many days, and apply only to the available time. When the error performance is consistently bad (a BER worse than 10^{-3} in ten consecutive one-sec-

Table 5.2 Error performance objectives; an example (CCITT, 1988a)

Performance parameter	Objective
Degraded minutes	Fewer than 10% of one-minute intervals to have a BER worse than 10^{-6}
Severely errored seconds	Fewer than 0.2% of one-second intervals to have a BER worse than 10^{-3}
Errored seconds	Fewer than 8% of one-second intervals to have any errors (i.e. EFS > 92%)

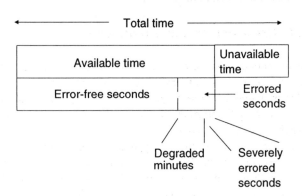

Figure 5.9 Relationship between error performance parameters

ond intervals is specified) the connection is deemed to be unavailable for information transmission. Figure 5.9 illustrates the relationship between the various performance parameters. Some subtleties of the terminology are to be found in the full specification (CCITT, 1988a).

From the network operators point of view, this specification is somewhat incomplete. With the increasing deployment of digital networks a connection will often comprise a number of links perhaps of very different type. The network operator needs to consider the effects of these individual links on the overall G.821 objectives which the customer expects. Outline Recommendation M.550 (CCITT, 1988b) attempts to address these matters by defining performance objectives for:

1. The long-term (several days); for bringing a digital link into service.
2. The short-term (several minutes/hours); for removing an unsatisfactory link for maintenance.

However, this is a complex and challenging topic which remains the subject of continuing study, such as Yamamoto and Wright (1989) and Kubat and Bollen (1989).

5.2.6 Error performance monitoring

For many applications it is considered highly desirable to monitor the occurrence of transmission errors while a system is in service. This enables problems to be seen quickly and appropriate action taken such as reducing the transmission rate to ensure better performance or switching to an alternative link.

Errors are monitored by adding redundancy to the information signal. This added redundancy may also be used for other purposes (such as with line coding and frame alignment) or specifically for error detection (as with error control coding). These topics are dealt with in more detail in following sections, and are briefly introduced below.

5.2.6.1 *Line coding*

Codes such as HDB3, 7B8B and the Manchester code can provide a monitoring capability on a per link basis. These codes have well defined coding rules which can be verified at a regenerator or subsequent link terminating point. Once any requisite code word synchronism has been obtained, any violations of the coding rules can be attributed to transmission errors and so the error performance of the link can be estimated.

5.2.6.2 *Frame alignment*

Here redundancy is added for synchronisation purposes. Once correct alignment is assured, any variation from the known frame alignment word can again be attributed to transmission errors. As with line codes, care has to be taken to distinguish between word misalignment conditions and transmission errors. The decision algorithms are often quite complicated but nevertheless can give a reasonable estimate of transmission error performance. Frame alignment words usually remain associated with the transmitted information over more than a single link and so can offer the opportunity for error monitoring over a greater portion of a connection.

5.2.6.3 *Error control coding*

It may not always be viable to provide an end-to-end level of error performance which is compatible with the most demanding digital service likely to be supported. In such circumstances error control coding (detection and then correction) over certain problem links or end-to-end over the whole connection may be used. This is the most direct method of error performance monitoring as the redundancy will usually have been specifically added, and so optimised, for the detection of errors. Of course certain error combinations may go undetected or only partially detected and this can lead to erroneous action in a subsequent error corrector. However, this problem can usually be reduced to an acceptable level by using coding which is appropriate to the error characteristics of the particular transmission link. It then becomes possible to ensure that transmission errors in blocks of data for example, are monitored to a high degree of confidence.

5.3 Line codes

5.3.1 Definition

A line code defines the equivalence between sets of digits generated in a terminal and the corresponding sequence of symbol elements

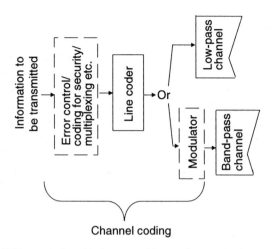

Channel coding

Figure 5.10 Relationship between line coding and other channel coding functions

transmitted over a channel. It must be chosen to suit the characteristics of the particular channel (Cattermole, 1983; Waters, 1983).

For metallic cables which have an essentially low-pass frequency characteristic it will usually be the last coding function performed before transmission. For band-pass channels such as optical fibres, radio and the analogue telephone network, where modulation is required before transmission, the line coding function will either immediately precede the modulator or be incorporated within the modulation process.

The relationship between line coding and other channel coding functions is shown in Figure 5.10. From this we see that the line coder may have to operate at the transmitted symbol rate, in which case a reasonably simple design will usually be essential, especially for high speed systems.

5.3.2 Purpose of a line code

A line code provides the transmitted symbol sequence with the necessary properties to ensure reliable transmission and subsequent detection at the receiving terminal. In order to achieve this the following have to be considered:

1. Spectrum at low frequency. Most transmission systems are easier to design if a.c. coupling is used between stages. This means that decoding must not rely on receiving a d.c. component. Furthermore, the coupling components (capacitors or transformers) will introduce a low frequency cut-off and this will put a limit on the permissible low frequency content of the line code if long term intersymbol interference is to be avoided (Cattermole, 1983).

2. Transmission bandwidth required. If transmission bandwidth is at a premium, a multi-level line code, in which each transmitted symbol represents more than one bit of information, will enable the transmitted symbol rate to be reduced. However, for a given bit error ratio, a multi-level line code will require a better signal-to-noise ratio than a corresponding binary code.

3. Timing content. There must always be sufficient embedded timing information in the transmitted symbol sequence. This will ensure that the distant receiver (and intermediate repeaters if used) can extract a reliable clock to time its decision making processes. This usually means ensuring that the line code provides an adequate density of transitions in the transmitted sequence.

4. Error monitoring. By adding redundancy into the information stream a line code can provide a means of in-service monitoring of the error rate of a transmission link. For example, a line coder can be constrained so that it never produces certain symbol sequences. Thus, the occurrence of these sequences at the receiver provides a means of estimating the link error performance.

5. Efficiency. In order to provide the above features it will usually be necessary to add extra information (redundancy as far as the data is concerned) into the digit stream. This will lower the efficiency of the line code which can be defined as the ratio of the average information carried per transmitted

symbol to the maximum possible information per symbol (when assuming no added redundancy), expressed as a percentage.

5.3.3 Classification of line codes

Line codes can be classified in a variety of ways, one method is to identify the following categories:

1. Bit-by-bit codes.
2. Block codes: bit insertion and block substitution.
3. Partial response codes.

Some codes will fall into more than one of these classes. The following are representative examples from each category.

5.3.3.1 *Manchester code*

This is also called diphase or WAL1 code and is a simple example of a bit-by-bit code. Each information bit is coded into a two-bit symbol for transmission, as shown in Figure 5.11(a). No d.c. component is

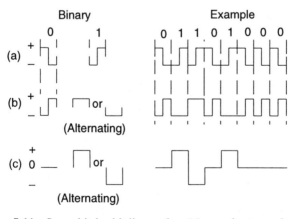

Figure 5.11 Some bit-by-bit line codes: (a) manchester code; (b) code mark inversion (CIM); (c) alternate mark inversion (AMI)

generated because the polarity and amplitude of the first half of each transmitted symbol is complemented by the second half. Also, the resulting signal transition in the centre of each symbol ensures an adequate clock content.

The absence of such a transition indicates a transmission error and so error monitoring is also possible. However, only one bit of information is carried for every two bits transmitted and so the efficiency is only 50%. In this case, this results in a channel bandwidth requirement of twice the uncoded bandwidth.

However, its simplicity makes it attractive for applications where transmission bandwidth considerations are less crucial; it is used in the magnetic recording of digital signals and in the Ethernet local area network system (ANSI/IEEE 802-3)

5.3.3.2 *Code mark inversion (CMI)*

It is shown in Figure 5.11(b) and is similar to the Manchester code except that it achieves a polarity balance over a longer term by alternating between a positive symbol and a negative symbol when a binary one (mark) is to be transmitted.

It is specified by ITU-T (CCITT, 1988c) as an interface code for short distance connections between certain types of transmission equipments.

5.3.3.3 *Alternate Mark Inversion (AMI).*

This differs from the above line codes in that the transmitted symbol rate is the same as the binary information rate. It provides the desired features by using a redundant signal set with three possible amplitude levels as shown in Figure 5.11(c).

The zero d.c. property is achieved by sending marks as a positive or negative symbol alternately. An estimate of the link error performance is obtained by counting violations of this alternating mark inversion rule.

We note that AMI has a potential weakness; if many successive zero-level symbols are transmitted, then the clock will fail. Thus, either the number of successive zeros has to be restricted, which is undesirable for a transparent data link, or a code substitution has to be made to prevent this potential hazard. Such AMI based substitution codes are used in 24 and 30 channel PCM telephone systems, where they are known as B6ZS and HDB3 respectively (Waters, 1983).

5.3.3.4 mB1C

This (Yoshikai et al., 1984) is an example of a block code of the bit insertion type. A block of m information bits has added to it a single extra bit which is selected in an attempt to provide a long term polarity balance and so no d.c. component. For this to work reliably, the source data may need to be scrambled first to ensure that it is adequately randomised. The added bit also ensures a minimum clock content and some redundancy to permit transmission error monitoring. Binary codes of this type feature high efficiency (m has ranged in practice from 7 to at least 23) and simple hardware, and they are thus attractive for high speed applications such as in optical fibre systems. Codes such as mB1P (Dawson and Kitchen, 1986) and DmB1M (Kawanishi et. al., 1988) are similar in concept but offer slightly different features.

5.3.3.5 mBnB

These (Brooks and Jessop, 1983) are block codes where m binary source bits are mapped into n binary bits for transmission. Redundancy is built into the code to provide the desired transmission features by making n > m. Several such codes have been proposed (and used), in particular where n = m + 1. They differ from the previous category in that the m-bit transmitted block may bear little similarity to its input source block. This gives greater flexibility for providing the desired line code features for a given level code efficiency, but is achieved at the expense of increased encoder and decoder circuit complexity.

By way of example, part of the translation table for the balanced polarity 7B8B code (Sharland and Stevenson, 1983) is illustrated in Table 5.3. Each 7-bit input source word is mapped into one or two possible 8-bit output words depending on the polarity balance, or

Table 5.3   Part of the translation table for 7B8B

Input (7 bits)	Output as transmitted (8 bits) Word disparity		
	Negative	Zero	Positive
1111111		− + − + − + − +	
1010001	+ − − − + − − +	or	− + + + − + + −
0101001	− + − − − + − −	or	+ − + + + − + +

disparity, of the transmitted words. Output words which are balanced in themselves (that is, have zero disparity) are to be preferred in that a single input-to-output mapping is sufficient. However, there will only be a limited number of these and so the table has to be completed by using non-zero disparity pairs where members of each pair have a disparity of opposite sign. The selection of output words is then made on the basis of minimising the cumulative disparity. It follows that some possible output words will not be needed, this redundancy provides the necessary design flexibility. The selection of words and their mappings from input-to-output is made on the basis of ensuring: good timing content, error monitoring, word alignment and of minimising the opportunity for transmission error multiplication in the decoding process. A computer search may be used to optimise this mapping.

Many such binary-to-binary balanced codes have been designed. A similar design philosophy can also be used for bandwidth efficient multi-level codes, of which 4B3T (4 binary bits converted to 3 ternary symbols) (Catchpole, 1975) and subsequent variations on it are perhaps the best known.

5.3.3.6 *Partial response*

These codes (Lender, 1981), known also as correlative or multiple response codes, deliberately introduce a controlled amount of intersymbol interference into the received signal. This known amount of correlated interference produces a multi-level received signal which, with suitable pre-transmit coding, can be designed to provide the requisite line code features.

A simple example of the class is duobinary (Lender, 1966) in which a binary signal is transmitted at twice the rate required to give zero intersymbol interference. With suitable coding and equalisation the resulting intersymbol interference can be arranged to give a 3-level received signal.

Figure 5.12 gives waveforms for this code. Figure 5.12(a) shows an input step applied to a transmission channel which, after appropriate equalisation, produces the requisite output at the receiver after time T (ignoring channel delay) Figure 5.12(b).

Now, if signals are launched into the channel with a time spacing of less than T, the output will have time to only partially respond to the input stimuli. Figure 5.12(c) shows such an input where the symbol period is only T/2, that is, signalling is at twice the conventional rate.

With duobinary, some pre-transmit coding is used as detailed in Figure 5.12(d), the waveform being shown in Figure 5.12(e). This signal is launched into the channel and its associated equaliser, the result is shown in Figure 5.12(f).

This received signal can then be applied to a 3 level digital decision circuit and decoded in accordance with the rules given in Figure 5.12(g) to yield an output which is the same as the original binary input.

Thus, a binary data signal has been transmitted at twice the conventional rate, the price paid being a 3-level signal at the receiver. The code features adequate transitions for clock extraction and also some redundancy suitable for transmission error monitoring, by noting that the output cannot change between the two outer levels in adjacent decision times.

For well behaved channels (where intersymbol interference can be accurately controlled), partial response codes can provide a bandwidth efficient transmission technique. They have been used in radio systems and also to convert existing 24-channel PCM systems into 48-channel systems with a minimum of equipment change.

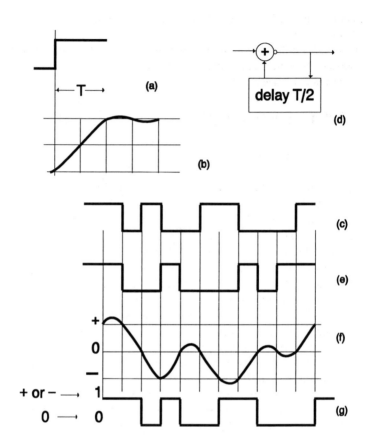

Figure 5.12 Duobinary, a partial response code: (a) step input to channel; (b) channel equaliser response; (c) binary input; (d) pre-transmit coding; (e) waveform for pre-transmit coding; (f) received signal; (g) decoded output

5.3.3.7 *Other line codes*

Many line codes, designed to meet the requirements of a variety of transmission systems are known. Useful surveys, together with references to further details, are to be found in for example Duc and Smith

(1977), International J. of Electronics (1983), Bylanski and Ingram (1987).

A rather different approach suitable for high bit rate systems which currently require simple encoding and decoding procedures is to be found in Fair et. al. (1991).

5.4 Clock extraction

5.4.1 Self-timed systems

Clock extraction, (known also as timing recovery or bit synchronisation), is the process by which a digital regenerator obtains a synchronising signal which enables it to optimise the timing of its decision making process. (Decision making and its relationship to the eye of the received signal has been discussed elsewhere.)

Most digital transmission systems are self-timed in that they extract the clock from the incoming data stream. This avoids the need for a separate timing channel, as used in most computers for example, and makes it easier over the longer transmission distances involved, to maintain the crucial phase relationship between the data and the clock at the point of decision making.

Self-timing requires that the data signal is coded to ensure that there is either a clock component present in the transmitted signal or that such a component can be reliably extracted after processing at the receiver. The words reliably extracted imply that a satisfactory clock can be recovered which has a frequency, phase and amplitude which is adequately immune from transmission distortions and interference, and from the effects of different data sequences. The latter requirement is found to be especially important in transmission systems containing a number of regenerative repeaters in tandem. Each repeater will experience some pattern dependent variation or jitter on the phase of its extracted clock and this will be passed on via the regenerated data stream to the next repeater. Thus the locally generated phase jitter will combine with the incoming signal jitter, a situation which will lead to a progressive build-up of jitter at subsequent regenerators. Careful design is required to ensure that this potential timing problem remains within acceptable bounds.

5.4.2 Timing content of a digital signal

All pulse sequences have symmetries in the amplitude and phase of their spectral structure. Even if the received signal lacks a component at the clock frequency, its spectral symmetry, which is independent of sequence statistics can, after further processing, be used to provide a clock for retiming purposes.

Analysis (O'Reilly, 1984) reveals that any repetitive impulse sequence, including the important case where the sequence length tends to infinity, will possess a frequency spectrum with even amplitude symmetry and odd phase symmetry (that is, Hermitian symmetry) about half the clock frequency. These symmetry conditions, which are independent of sequence statistics, arise from the periodic structure associated with the signal clock.

Now, if a received sequence has little or no d.c. component, it follows directly from the above symmetry property that it will also have little or no f_c component. Most transmission codes, which deliberately aim to suppress any d.c. content in a data sequence will fall into this category. For such sequences further processing is required before a clock component can be extracted. This can be achieved by subjecting the received signal to a non-linearity such as rectification, clipping or squaring or, often in practice, a combination of these. The Hermitian symmetry within the received signal, when subjected to the non-linearity results in pairs of spectral components, at frequency f_c, with equal amplitude and complementary phase. These components combine in a constructive manner to produce a non-zero spectral line at frequency f_c (Blackman and Mousavinez-had, 1990).

5.4.3 Clock extraction circuits

A commonly used method of clock extraction for data signals which have little or no spectral energy at the clock frequency is shown in Figure 5.13. The previous section has discussed how a signal spectrum with Hermitian symmetry about $f_c/2$ will, after non-linear processing, produce a spectral component at the clock frequency. Ideally, to ensure this symmetry property, it is necessary to have a separate

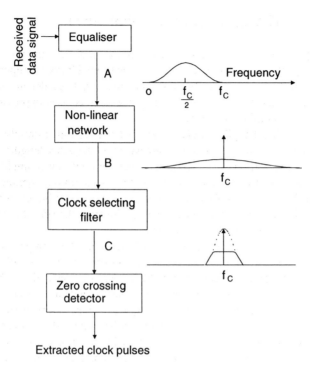

Figure 5.13 Clock extraction using a non-linearity

equaliser for the timing path. In practice, the equaliser of Figure 5.13 will often be optimised for the decision path and provide little more than noise bandlimiting for the timing path. Nevertheless, a Hermitian symmetry component will be present at point A even if it is diluted by the characteristics of the transmitter and channel.

The non-linear network then produces, at point B, a discrete spectral line at the clock frequency f_c together with a continuous spectrum arising from the random data pattern. The purpose of the clock selecting filter is to select the wanted clock whilst rejecting as much as possible of the pattern related spectrum together with any associated noise. The selection filter may be a narrow bandpass filter or a phase locked loop with effective Q factors ranging in practice from about 80 to in excess of 200. The choice of Q will depend critically upon the stability of the incoming data rate and the tuning accuracy of the filter. A high Q will minimise the unwanted spectral components but will give rise to large phase shifts in the extracted clock if the tuned frequency deviates from the incoming clock frequency f_c.

Finally, the signal at point C is applied to a zero crossing detector or hard limiter which removes the amplitude variations and give clock pulses for use in the decision making and regenerating circuits.

5.4.4 Jitter

Short term variations from the optimum timing of a digital decision making process are known as jitter. It can be regarded as a phase modulation of the extracted clock relative to the original system clock and as such is sometimes referred to as phase jitter. The effect of jitter on the decision making process within a regenerator is illustrated in Figure 5.14(a), where it can be seen that clock edge jitter can cause decisions to be made at sub-optimum times, thereby increasing the probability of error. This jitter is caused by a combination of:

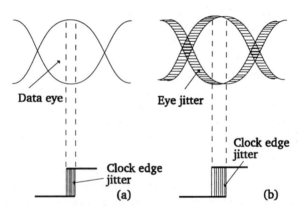

Figure 5.14 Timing jitter and the data eye: (a) clock edge jitter; (b) effect of jitter accumulation

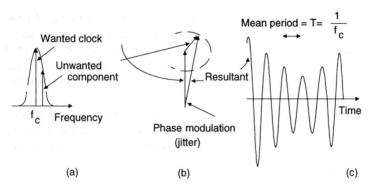

Figure 5.15 Jitter arising from an interfering spectral component: (a) spectrum from clock selecting filter (Figure 5.13 point C); (b) phasor representation of (a); (c) resulting waveform with amplitude and phase modulation

1. Incoming noise affecting the extraction process.
2. Data dependent timing effects arising from imperfections and limitations in the clock extraction process.

Both of these effects arise from the unwanted continuous spectral components getting through the clock selecting filter to point C in Figure 5.13. They combine with the spectral component of the wanted incoming clock to produce a resultant output clock which exhibits both amplitude and phase modulation. The mechanism for a single interfering component is shown by way of example in Figure 5.15. Figure 5.15(a) shows the wanted and unwanted spectral components as they might appear at the output of a clock selecting filter. The equivalent phasor representation is given in Figure 5.15(b) where it is seen that the resultant has both amplitude and phase modulation. This is again apparent in the corresponding waveform of Figure 5.15(c). The amplitude variations can be effectively removed by a zero-crossing detector provided that the clock component is always adequately large. However, the phase modulation will remain giving rise to timing jitter.

In practice appropriate noise bandlimiting at the regenerator input means that the contribution to jitter from the noise is usually small, that is, the data dependent timing effects tend to dominate. These effects can be removed if the output spectrum of the clock selecting filter also has Hermitian symmetry, but which is now centred on f_c (Franks and Bubrowski, 1974). This can be explained by considering the phasor diagram of Figure 5.15(b). If the unwanted components always had Hermitian symmetry centred on f_c each pair would produce a vector of equal amplitude but opposite phase with respect to the unwanted clock. These would rotate about the f_c vector in opposite directions and so would produce a resultant which had amplitude modulation but no phase modulation, that is, no jitter.

It has already been noted that any data sequence (in impulse form) will have Hermitian symmetry properties which, after squaring, will produce the requisite Hermitian symmetry about f_c. It thus remains to provide equalisation to compensate for finite width and channel characteristics to ensure that the spectral symmetry condition is preserved at the output of the clock selecting filter. Unfortunately this requirement is not consistent with the skew symmetry spectral requirements for a good received eye and so must be provided by separate spectral shaping within the clock path. Although this demonstrates a theoretical requirement, in practice separate clock path equalisation is not always found to be necessary. Acceptably low jitter performance can often be obtained by a judicious combination of: careful transmission coding, to ensure an adequate clock component relative to the continuous spectrum after non-linear processing (Jones and Zhu, 1987), received noise limiting, and an appropriate choice of Q for the clock selecting filter.

5.4.5 Jitter accumulation

Jitter arising from the clock extraction and retiming process in a regenerator will be transmitted to subsequent regenerators where it may combine with any locally generated timing variations to cause an overall accumulation of jitter. When assessing the effect of this jitter accumulation in a self-timed system, two inter-related factors have to be taken into account. These are eye jitter and clock edge jitter and are depicted in Figure 5.14(b). Eye jitter is a direct consequence of any timing variations already present on the incoming data signal.

Clock edge jitter arises as described in the previous section (which refers to Figure 5.14(a)) except that now a contribution will also be made to it by the incoming data signal.

The decision making process at the regenerator may not be as difficult as implied by Figure 5.14(b). The extracted clock will tend to follow any phase variations in the incoming data. What matters as far as the instantaneous signal-to-noise ratio and so decision error probability is concerned, is the relative timing of a particular decision instant. That is, the difference between the time position at the clock edge with respect to the optimum decision time for the particular incoming data symbol. Although this jitter tracking may help the decision making process at a particular regenerator, jitter will still accumulate from one regenerator to another. Problems can arise at the end of a link where the data signal may have to relate to a more rigid network clock. At this point, the incoming data signal may contain timing jitter which amounts to several clock periods. If data slips (loss or gain of bits) are to be avoided quite large data buffers may be required. The way in which jitter builds up has been extensively studied in the literature (Byrne et al., 1963). Suffice it to say here that it is the data pattern dependent effects which cause most trouble as they tend to reinforce each other at subsequent repeaters. In long chains of regenerators it is usually preferable to prevent the build up of excessive jitter in the first place by using carefully designed phase locked loops rather than band-pass filters in the clock selection circuits. Alternatively, data scramblers may be installed at intervals to break up the data patterns and so control the main contributor to the problem, namely, the build-up of data dependent jitter.

The amount of jitter passed through a regenerator will depend upon the detail of the clock extraction process and in particular on the parameters of the clock selecting filter. For a simple tuned circuit the jitter power transfer characteristic (Byrne et al., 1963) will depend upon the Q of the filter as shown in Figure 5.16. Thus, low frequency jitter will be transferred through a regenerator with very little attenuation whilst higher frequency jitter will be reduced. It follows that any jitter suppressors or data buffers will need to be designed to deal especially with low frequency jitter, thus they may need to store large numbers of consecutive data bits so that slow timing variations can be smoothed out.

5.4.6 Jitter specifications

Specifications for jitter accumulation in digital systems have to take account of the low frequency build-up implied in Figure 5.16. A typical specification for jitter limits for a digital transmission system comprising many regenerators is given in Figure 5.17 where peak-to-peak jitter, expressed in clock periods, is plotted against jitter frequency. (It should be noted that r.m.s. values will be typically a factor of ten less than these figures.)

This representative example is loosely based on figures given in ITU-T Recommendations G823 and G824, although reference to these will show that the higher multiplex levels in the transmission hierarchy attract somewhat tighter specifications. The figure serves to highlight the low frequency timing jitter (or wander) which can amount to many clock intervals.

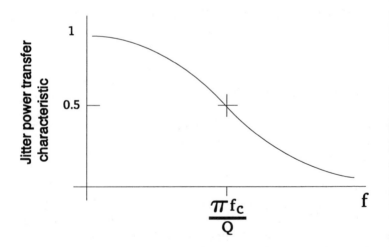

Figure 5.16 Jitter transfer characteristics for a tuned circuit

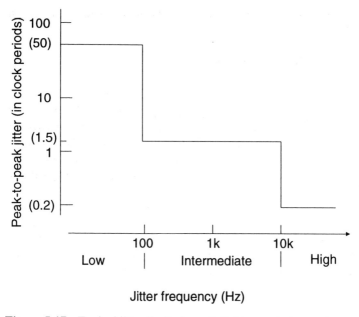

Figure 5.17 Typical jitter limits for a digital transmission system

5.4.7 Further reading

Apart from the references given in the text, useful chapters on clock extraction are to be found in Bylanski and Ingram (1987) and also in (Lee and Messerschmitt, 1988), the latter includes details of waveform sampling methods which are applicable to fully digital realisations.

For detailed analytical treatment the early work on timing and jitter published in the Bell System Technical Journal should be consulted (in particular the papers by Bennett and the one by Rowe, all in Vol 37, 1958 and also the paper by Manley in Vol 48, 1969).

On jitter in digital networks, (Kearsey and McLintock, 1984) is a helpful paper which relates closely to the ITU-T recommendations.

5.5 Frame alignment

5.5.1 The task

In addition to clock extraction, most signal formats used for digital transmission will entail some further structure which must be reliably extracted at the receiving terminal. In particular, when several information streams are combined before transmission using time division multiplexing, a frame structure will be involved. The boundaries between frames are usually marked by inserting a carefully selected frame alignment word (also called a marker, flag or frame sync pattern) which the receiver has to locate before demultiplexing can be performed (Figure 5.18).

The reliable frame alignment of such multiplexed signals is essential for the proper functioning of many digital communication systems. A loss of alignment will result in the failure to correctly identify the received bits and so will cause the disorientation of the demultiplexing process. This leads to a catastrophic loss of both message and control information. Thus the choice of frame alignment word (FAW) and the design of reliable alignment detection, misalignment detection and searching algorithms is crucial. Figure 5.19 shows the principal tasks which have to be performed, with the desirable state shown shaded.

In practice, the aligner's task is made yet more challenging when one notes that:

1. The frame alignment word may, in most systems, be temporarily imitated in random positions within the data region.

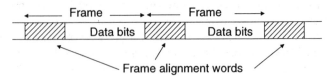

Figure 5.18   Frame structure

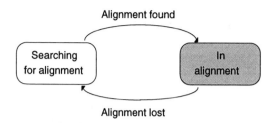

Figure 5.19   The alignment task

2.   The frame alignment word may be mutilated by transmission errors.

3.   In some systems bit slips can occur, that is, frames may temporarily lose or acquire erroneous bits.

To deal with these problems the 2-state model of Figure 5.19 has to be extended to at least the 3-state model of Figure 5.20. This model distinguishes between two locked (not searching) states; the desirable in alignment and locked state and the highly undesirable out of alignment but locked state. In this latter state a misalignment condition has occurred but the system has not yet detected it. The causes for transition from one state to another are given in the diagram and it can be seen that they relate directly to the conditions (1) to (3) listed above.

Before considering the mechanisms involved in such an alignment process there are two variations to the above which can be found in practice. The FAW may not be transmitted in the bunched format shown in Figure 5.18. If a system is known to be prone to short bursts of transmission errors, some protection against them may be afforded by distributing the FAW in known positions throughout the frame. Whichever method is adopted, the alignment principles are similar. Secondly, some systems use a unique FAW, that is, the data bits are encoded so that in the absence of transmission errors the FAW can never be imitated (e.g. the frame marker flag 01111110 in ITU-T Recommendation X.25). For a given level of confidence in the alignment process, this will simplify the requisite algorithms but at the expense of some redundancy in coding the data bits.

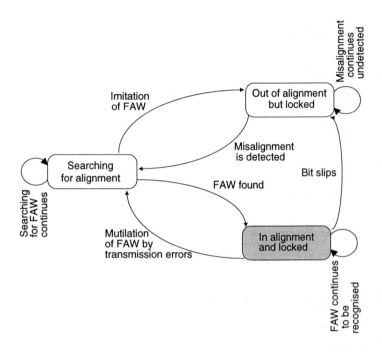

Figure 5.20 A state transition diagram for frame alignment

5.5.2 The alignment process

When a system is operating correctly the aligner should reside in the in alignment and locked state of Figure 5.20, with the equipment verifying that the FAW occurs in the predicted position in each frame. If a misalignment condition is then detected, the mechanism must transfer to the searching state and begin an inspection of all possible FAW positions. When the correct position has been found the search can be abandoned and the equipment allowed to revert to the locked condition.

In practice the alignment mechanism needs to be yet more complicated than that shown in Figure 5.20. This is because it is desirable to try and verify that the current state really is incorrect before transferring to another state. For example, when the equipment is in alignment and locked, transmission errors may occasionally mutilate the FAW in a given frame; it may be sensible to wait and check the next frame (or the next few frames) to minimise the probability of setting off on an unnecessary search. Similarly, when in the searching state, a match to the FAW may be found which in fact is an imitation caused by the data bits. This condition is unlikely to be duplicated in the same position in successive frames (unless the FAW chosen is a bad one!) and so again, a check on subsequent frames may be sensible before reverting to the locked condition. Thus in practice, a series of check states are inserted between the locked and searching states of Figure 5.20 to reduce the probability of erroneous transitions between states.

Of course a compromise is necessary, if a genuine transition to another state is required, it is equally undesirable to waste time making unnecessary checks. Unfortunately the equipment cannot instantaneously distinguish between necessary and erroneous transitions. Table 5.4 gives some examples of practical frame alignment arrangements, the last two columns detail the number of frame checks recommended before changing state.

To summarise, when in alignment, a good frame alignment process should exhibit a low probability of losing alignment and, when, out of alignment, a high probability of fast recovery to the aligned condition. To achieve these objectives the following features are required:

1. Reliable verification of the in alignment condition.
2. An efficient search procedure which ensures rapid location and verification of the position of the FAW.
3. Rapid detection of a genuine out of alignment condition caused by bit slips.
4. Robust performance in the presence of FAW imitations and transmission errors.

Ideally these features should be obtained with a minimum of added redundancy in the data stream (in the form of an FAW) and with reasonably simple alignment equipment.

Table 5.4   ITU-T Recommendations for frame alignment, for the 30 channel multiplexer hierarchy

Multiplexer level	ITU-T Rec.	Frame length (bits)	Frame alignment word	Number of frame checks before changing state	
				Lock-to-search	Search-to-lock
Primary (approx. 2Mbits/s)	G732	512	0011011 (7 bits)	3 or 4	3
Second (approx. 8Mbits/s)	G745	1056	11100110 (8 bits)	5	2
Third (approx. 34Mbits/s)	G751	1536	1111010000 (10 bits)	4	3
Fourth (approx. 140Mbits/s)	G751	2928	111110100000 (12 bits)	4	3

5.5.3 Searching techniques

A simple bit-by-bit aligner is outlined in Figure 5.21. A test window, of length equal to the FAW, is shown to be correctly aligned with the incoming signal. That is, a comparator (or pattern matcher) indicates that there is agreement between the received FAW and a locally stored version. This causes a bit counter to count through the known frame length and so locate and verify the position of the next FAW, and so on. If an FAW is not found in the expected position the control circuit will initiate a bit-by-bit search through the frame until a new alignment position is found.

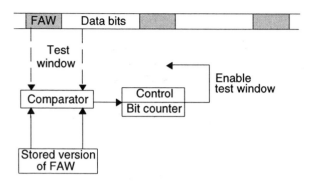

Figure 5.21   Basic configuration for a frame aligner

As discussed in the previous section, it will usually also be necessary to guard against erroneous action being taken when an FAW is mutilated by transmission errors or when an imitation of an FAW occurs within the data region. We have seen that this can be done by checking an appropriate number of frames before entering or leaving the search state. A further sophistication is that of off-line searching in which extra equipment commences a search as soon as a misalignment condition is suspected. This takes place in parallel with the normal receiver processing and offers the considerable advantage that if the search proves to have been unnecessary no damage will have been done. It also benefits the true out of alignment condition in that an early start will have been made on the searching process.

The bit-by-bit searching technique can be relatively slow in establishing the aligned condition when there are random FAW imitations within the data region. This arises because each time an imitation is found the search will stop and a check made for an FAW one frame later. On failing to find the requisite pattern, the search will resume. This temporary halt in the searching process will occur every time an FAW imitation is found. A faster search can be achieved by using more memory. For example, the location of sequences which match the FAW can be simply recorded and the search allowed to continue. Those which do not repeat at one-frame intervals are ignored. This process continues until the location of the true FAW becomes apparent. The aligner can then be made to jump directly to this new position. Thus, there is a trade-off; faster searches are possible but at the expense of equipment complexity.

5.5.4 Choice of frame alignment word

Some recommended patterns have been given in Table 5.4. By way of example, Figure 5.22 takes the first pattern in the Table (for 30-channel pcm) and plots the mean number of bit matches from the alignment comparator for different window positions. When the test window is in the correct position (and there are no transmission errors) the comparator will register seven agreements. If the position of the window is in error by 1 bit (in either direction), then it can be seen that three agreements and three disagreements will be registered, with the seventh bit overlapping into the data region. For random data this last bit will register an agreement or disagreement with equal probability, thus overall, the mean number of agreements for a window position error of 1-bit will be 3.5. Repeating this calculation for other window positions generates the plot of Figure 5.22. This confirms that the sequence 0011011 is a good FAW in that it produces a strong correlation spike when in alignment with relatively low sidelobes associated with the out of alignment window positions. In this respect we note that the correlation values for alignment errors of 1 or 2 bits are especially low, that is, it is a good word to use for well behaved systems which rarely experience more than 1 or 2 bit slips. Under such circumstances the pattern is thus seen to be reasonably immune to unfavourable data bit combinations and/or isolated transmission errors, both of which could increase the apparent number of agreements but are unlikely to achieve the in alignment peak value.

The choice of FAW for a particular system will depend upon many factors. For example, in contrast to the above well behaved application, systems which have to acquire alignment frequently, particularly if from a random starting position, will usually require longer FAWs. On the other hand, long FAWs are more liable to experience transmission errors. Thus, the word choice and the alignment strategy are closely linked. For example, in the presence of transmission errors it may be worthwhile to accept window positions that demonstrate a

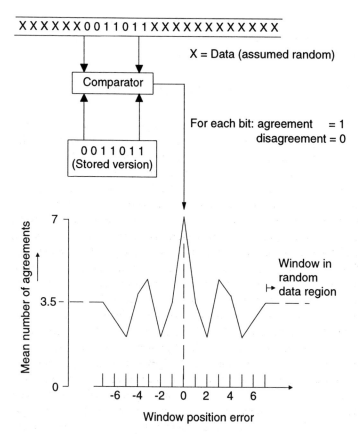

Figure 5.22 Mean number of bit agreements for different window positions

close but not necessarily exact match to the expected FAW. The implications of this and other variations on the basic alignment approach described above are to be found in the literature.

5.5.5 Further reading

A good introduction to frame alignment is to be found in (Bylanski and Ingram, 1986) and in several more recent books which refer to them and adopt a similar approach. Some of the subtleties of the alignment process with examples based on ITU-T specifications are discussed in (Jones and Zhu, 1985), a similar method of analysis (using probability generating functions) is adopted in a tutorial article (Choi, 1990). Barker (Barker, 1953) was first to study the selection of good frame alignment words, since then many variations have been investigated to provide for optimum performance under a variety of system conditions, a recent study with references to earlier work is to be found in (Al-Subbagh and Jones, 1988).

5.6 References

Al-Subbagh, M.N. and Jones, E.V. (1988) Optimum Patterns for Frame Alignment, *Proceedings IEE (Part F)*, **135**, pp. 594-603.

Barker, R.H. (1953) Group Synchronising of Binary Digital Systems. In *Communication Theory*, (ed. W. Jackson), Academic Press, pp. 273-287.

Beaulieu, N.C. (1989) A Simple Series for Personal Computer Computation of the Error Function, *IEEE Transactions on Communications*, **3**, pp. 989-991, (September).

Bennett, W.R. and Davey, J.R. (1965) *Data Transmission*, McGraw-Hill.

Blackman, N.M. and Mousavinezhad, S.H. (1990) The Spectrum of the Square of a Synchronous Random Pulse Train, *IEEE Transactions on Communications*, **38**, pp. 13-17.

Brooks R.M. and Jessop, A. (1983) Line Coding for Optical Fibre Systems, *International Journal of Electronics*, **55**, pp. 81-120.

Bylanski, P. and Ingram, D.G.W. (1987) *Digital Transmission Systems*, 2nd edn. Peter Peregrinus.

Byrne, C.J., Karafin, B.J. and Robinson, D.R. (1963) Systematic Jitter in a Chain of Digital Regenerators, *Bell System Technical Journal*, **42**, pp. 2679-2714.

Carlson, A.B. (1986) *Communication Systems*, 3rd edn., McGraw-Hill.

Catchpole, R.J. (1975) Efficient Ternary Transmission Codes, *Electronic Letters*, **11**, pp. 482-484, (October)

Cattermole, K.W. (1983) Principles of Digital Line Coding, *International Journal of Electronics*, **55**, pp. 3-33.

CCIR (1986) Report 967-1, *Digital Television: Transmission Impairments and Methods of Protection.*

CCITT (1988a) Recommendation G.821, *Error Performance of an International Digital Connection Forming Part of an Integrated Services Digital Network*, Blue Books, Geneva.

CCITT (1988b) Recommendation M.550. *Performance Limits for Bringing into Service and Maintenance of Digital Paths, Sections and Line Sections*, Blue Books, Geneva.

CCITT (1988c) Recommendation G.703, *General Aspects of Interfaces*, Blue Books, Geneva.

Choi, D.W. (1990) Frame Alignment in a Digital Carrier System — a Tutorial, *IEEE Communications Magazine*, pp. 47-54 (February).

Dawson, P.A. and Kitchen J.A. (1986) TAT-8 Supervisory System, *British Telecom Engineering Journal*, **5**, (July).

Duc, N.Q. and Smith B.M. (1977) Line Coding for Digital Data Transmission, *Australia Telecommunications Research Journal*, **11**, pp. 14-27.

Fair, I.J., Grover, W.D., Krzymien, W.A. and MacDonald, R.I. (1991) Guided Scrambling: A New Line Coding Technique for High Bit Rate Fibre Optic Transmission Systems, *IEEE Transactions on Communications*, **39**, pp. 289-297 (February).

Franks, L.E. (1968) Further Results on Nyquist's Problem in Pulse Transmission, *IEEE Transactions on Communications*, **16**, pp. 337-340.

Franks, L.E. and Bubrowski J.P. (1974) Statistical Properties of Timing Jitter in a PAM Timing Recovery System, *IEEE Transactions on Communications*, **22**, pp. 913-920.

Hamming, R.W. (1980) *Coding and Information Theory*, Prentice-Hall.

International Journal of Electronics (1983) **55**, Special Issue on Line Codes.

Jones, E.V. and Al-Subbagh, M.N. (1985) Algorithms for Frame Alignment — Some Comparisons, *Proceedings IEE (Part F)*, **132**, pp. 529-536.

Jones, E.V. and Zhu, S. (1987) Data Sequence Coding for Low Jitter Timing Recovery, *Electronics Letters*, **23**, pp. 337-338.

Kanal, L.N. and Sastry, A.R.K. (1978) Models for Channels with Memory and their Applications to Error Control, *Proceedings IEEE*, **66**, pp. 724-744.

Kawanishi, S. et al. (1988) DmB1M code and its Performance in a Very High Speed Optical Transmission System, *IEEE Transactions on Communications*, **36**, pp. 951-956, (August).

Kearsey, B.N. and McLintock, R.W. (1984) Jitter in Digital Telecommunication Networks, *British Telecommunications Engineering*, **3**, pp. 108-116 (July).

Knowles, M.D. and Drukarev, A.L. (1988) Bit Error Rate Estimation for Channels with Memory, *IEEE Transactions on Communications*, **36**, pp. 767-769.

Kubat, P. and Bollen, R.E. (1989) A Digital Circuit Performance Analysis for Tandem Burst-Error Links in an ISDN Environment, *IEEE Transactions on Communications*, **37**, pp. 1071-1076.

Lee, E.A. and Messerschmitt, D.G. (1988) Digital Communication, Kluwer Academic.

Lender, A. (1966) Correlative Level Coding for Binary-Data Transmission, *IEEE Spectrum*, pp. 104-115 (February).

Lender, A. (1981) Correlative (Partial Response) Techniques. In *Digital Communications: Microwave Applications*, (ed. K. Feher), Chapter 7, Prentice Hall.

McLintock, R.W. and Kearsey, B.N. (1984) Error Performance Objectives for Digital Networks, *The Radio and Electronic Engineer*, **54**, pp. 79-85, (February). (A later version which includes changes in CCITT Recommendations is to be found in: *British Telecommunications Engineering*, **3**, pp 92-98, July 1984.)

Nyquist, H. (1928) Certain Topics in Telegraph Transmission Theory, *Transactions AIEE*, **47** pp. 617-644.

O'Reilly, J.J. (1984) Timing Extraction for Baseband Digital Transmission. In *Problems of Randomness in Communication Engineering* (ed. Cattermole and O'Reilly), Pentech Press.

Sharland, A.J. and Stevenson, A. (1983) A Simple In-Service Error Detection Scheme Based on the Statistical Properties of Line

Codes for Optical Fibre Systems, *International Journal of Electronics*, **55**, pp. 141-158.

Sklar, B. (1988) *Digital Communications*, Prentice-Hall.

Sunde, E.D. (1961) Pulse Transmission by AM, FM, and PM in the Presence of Phase Distortion, *Bell System Technical Journal*, **40**, pp. 353-422.

Takahashi, K. (1988) Transmission Quality of Evolving Telephone Services, *IEEE Communications Magazine*, pp. 24-35, (October).

Waters, D.B. (1983) Line Codes for Metallic Cable Systems. *International Journal of Electronics*, **55**, pp. 159-169.

Yamamoto, Y. and Wright, T. (1989) Error Performance in Evolving Digital Networks Including ISDNs, *IEEE Communications Magazine*, pp. 12-18 (April).

Yoshikai N.; Katagiri, K. and Ito, T. (1984) mB1C Code and its Performance in an Optical Communication System, *IEEE Transactions on Communications*, **32**, pp. 163-169.

6. Telecommunication cables

6.1 Introduction

Cables provide the transmission medium for the majority of telecommunication systems. The installed cost of cables is a significant proportion of total investment in any network. A fascinating aspect of telecommunication transmission history has been the achievement of progressive upgrades in traffic capacity over existing cable links. For example, the application of 30 channel systems to de-loaded audio pairs, the operation of coaxial cables initially with 300 channel (1.3MHz) systems and many years later with either 2700 channel (12MHz) systems or 1920 (140Mbit/s) digital systems.

Today, virtually all new cables except for the local area, are of optical fibre and of single mode design. Not only does single mode provide economic solutions to today's requirements, it also holds the greatest promise for exploitation as the bearer for high capacity systems yet to be designed.

This chapter gives a description of the various cables used in telecommunication networks and their impairments on signal transmission. Copper pairs, coaxial cables and various optical fibres are included.

6.2 Symmetric pair cables

Symmetric pair or balanced pair cables provide the most economic solution for the direct provision of audio circuits over short distances, for example from local exchange to street cabinets and distribution pillars. Notwithstanding new techniques in mobile radio and optical fibre, conventional copper pair cables for new communities, and for the enhancement and replacement of very old cables in established areas, continue to be in substantial demand.

A typical local area cable is of unit twin construction. The conductors are insulated with cellular polyethylene of various colours which provides for pair identification within the cable. Two insulated conductors are twisted together to form a pair, the twist length and that of the other pairs, being specially chosen to minimise crosstalk. The pairs are then assembled into basic units of either twenty-five, fifty or one hundred pair units. Each unit is identified by a lapping of coloured tape.

To prevent the cables filling with water in the event of damage to the cable sheath, or due to a faulty joint sheath closure, the cable can be fully filled with petroleum jelly during manufacture. The cable sheath is of polyethylene with an internal Glover barrier. This vapour barrier comprises an aluminium polyethylene laminate tape which is applied longitudinally with a small overlap, and with the polyethylene surface facing outwards. As the polyethylene sheath is extruded around the finished cable it bonds to the laminate. This APL sheath provides an effective moisture barrier and electrostatic screen.

Cables of this design can be supplied with conductor diameters ranging from 0.32mm to 0.9mm, and in sizes ranging from 100 pairs to 3200 pairs or more. The maximum average resistance and mutual capacitance is shown in Table 6.1. Overall cable diameters are in the

Table 6.1   Typical characteristics of unit twin cable

Conductor diameter (mm)	Maximum average resistance (ohms/km @ $20^{\circ}C$)	Maximum average mutual capacitance (nF/km)
0.32	223	56
0.40	143	56
0.50	91	56
0.63	58	56
0.90	28	56

Table 6.2 Typical characteristics of star quad cable

Wire diameter (mm)	Average mutual capacitance (nF/km)			Nominal resistance (ohm/km @ 10^0C)
	Nominal	Mini-mum	Maxi-mum	
0.63	45	38	49	53.2
0.9	41	39	44	26.0

range 17mm to 70mm, depending on conductor gauge and number of pairs. The capacitive unbalance at voice frequency is generally less than 275pF in a 500m length.

For longer circuits, cables with a lower mutual capacitance and less capacitance unbalance are required. These requirements are met with a star quad construction i.e. diagonally opposite conductors form a pair. A typical cable is formed from 0.63mm or 0.9mm cellular polyethylene insulated conductors, four being twisted into a star quad formation. The quads are then stranded into a concentric layered cable, the direction of standing alternating in successive layers. The complete cable of between 14 and 520 quads is then sheathed as described for the unit twin design. The resultant cable has a smaller diameter then the equivalent unit twin design and a more tightly toleranced characteristic as shown in Table 6.2. The capacitance unbalance at voice frequency in a 500m length is generally less than 40pF between pairs in the same quad, less than 25pF between pairs in adjacent quads and less than 150pF to earth.

6.2.1 Analysis of balanced pair cables

The characteristics of a cable can be defined in terms of its primary coefficients, where the distributed characteristics are represented by circuit elements lumped at unit lengths apart.

The primary coefficients of a uniform transmission line are:

R = Resistance in Ohms per unit length
G = Leakance in Siemens per unit length
L = Inductance in Henries per unit length
C = Capacitance in Farads per unit length

The secondary coefficients are the propagation coefficient P and the characteristic impedance Z_O given by Equations 6.1 and 6.2.

$$P = [\,(\,R + j\omega L\,)\,(\,G + j\omega C\,)\,]^{\frac{1}{2}} \tag{6.1}$$

$$Z_O = \left[\frac{(\,R + j\omega L\,)}{(\,G + j\omega C\,)}\right]^{\frac{1}{2}} \tag{6.2}$$

The propagation coefficient P is complex, as in Equation 6.3.

$$P = \alpha + j\beta \tag{6.3}$$

The real part α of the propagation coefficient thus gives the attenuation of the line (in nepers per unit length) and is the attenuation coefficient. The imaginary part β is called the phase coefficient.

The relationship between the attenuation and phase coefficients and the primary coefficients is given by Equation 6.4.

$$\tanh(\,\alpha + j\beta\,) = [\,(\,R + j\omega L\,)\,(\,G + j\omega C\,)]^{\frac{1}{2}} \tag{6.4}$$

By expanding real and imaginary parts, Equations 6.5 and 6.6 can be obtained.

$$\alpha = [\,\tfrac{1}{2}\,((\,R^2 + \omega^2 L^2\,)\,(\,G^2 + \omega^2 C^2\,))^{\frac{1}{2}} \\ + \tfrac{1}{2}\,(\,RG - \omega^2 LC\,)]^{\frac{1}{2}} \tag{6.5}$$

$$\beta = [\tfrac{1}{2}\,((\,R^2 + \omega^2 L^2\,)\,(\,G^2 + \omega^2 C^2\,))^{\frac{1}{2}} \\ - \tfrac{1}{2}\,(\,RG - \omega^2 LC\,)]^{\frac{1}{2}} \tag{6.6}$$

At low frequencies, $\omega L \ll R$ and $\omega C \ll G$ giving Equation 6.7.

$$\alpha = (RG)^{1/2} \tag{6.7}$$

At audio frequencies $\omega L \ll R$ but $\omega C \gg G$ giving Equation 6.8.

$$\alpha = (\tfrac{1}{2} \omega CR)^{1/2} \tag{6.8}$$

At very high frequencies $\omega L \gg R$ and $\omega C \gg G$ giving Equation 6.9.

$$\alpha = \tfrac{1}{2} \left(\frac{R}{Z_O} + GZ_O \right) \tag{6.9}$$

The value of Z_O is given by Equation 6.10.

$$Z_O = \left(\frac{L}{C} \right)^{1/2} \tag{6.10}$$

But, at high frequencies skin effect causes R to increase in proportion to the square root of the frequency, and so at high frequencies the line attenuation increases as the square root of the frequency.

6.2.2 Loaded lines

A special case arises if LG = RC. Substituting this condition in Equation 6.5, then the attenuation coefficient $\alpha = (RG)^{1/2}$ which at voice frequencies is independent of frequency. In a real cable, RC will be larger than LG and increasing G would increase the attenuation. However, L can be increased by inserting lumped inductance in each leg of a pair.

Loading a line in this way was an important technique for providing long distance audio circuits. However, the lumped inductance forms a low pass filter with cable capacity which limits the line bandwidth as illustrated in Figure 6.1.

In order to apply PCM systems to such cables the loading coils have to be removed. Hence the term 'deloaded audio'.

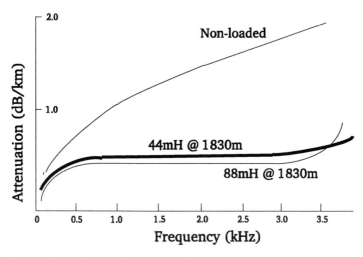

Figure 6.1 Effect of loading: 0.63mm conductors of mutual capacitance 45nF/km

6.2.3 Performance of balanced pair cables

Table 6.3 gives typical performance characteristics of 0.9mm star quad cable at audio frequencies. In this table 44mH/1800m indicates 44mH loading coils placed at 1800m. The line would only be operated at about 75% of the cut-off frequency. It will be seen that loading can reduce attenuation to one third of the unloaded value, but at the expense of restricted bandwidth.

Table 6.4 gives typical data for unit twin cable and shows how conductor size influences attenuation and impedance.

6.2.4 Pair cables for HF carrier systems

High frequency symmetric pair cables have been developed for analogue carrier telephone systems providing 12, 24, 36, 48, 60 and 120 channels on each pair, necessitating operation at up to 550kHz. These cables have polyethylene insulated pairs in star quad formation, and conductor sizes from 0.9mm to 1.3mm. A typical cable is

Table 6.3 Performance of 0.9mm star quad cable

Parameter	Non-loaded	44mH/1800m	88mH/1800m
Impedance @ 1kH	470Ω	820Ω	1130Ω
Attenuation @ 1kH	0.66dB/km	0.31dB/km	0.22dB/km
Cut-off frequency	—	5.7kHz	4kHz

Table 6.4 Performance of unloaded unit twin cable

Parameter	Conductor size			
	0.4mm	0.5mm	0.63mm	0.9mm
Impedance @ 1kHz (Ω)	910	725	567	400
Attenuation @ 1kHz (dB/km)	1.8	1.4	1.1	0.76

described in ITU-T (formerly CCITT) Recommendation G.611 (CCITT, 1989) from which Table 6.5 is an extract.

Far end crosstalk between pairs operating in the same direction has to be of the order of 70dB and this is achieved with the use of capacitive crosstalk balancing frames during cable installation.

6.2.5 Pair cables for digital transmission

As previously indicated, many 1.5Mbit/s and 2Mbit/s pcm systems have been provided over deloaded audio cables with regenerative repeaters equipped at the old loading coil sites. For this application the cable loss at 1MHz must not exceed about 30dB. However, the limiting factor on cable fill (the percentage of pairs in a given cable which can be equipped with pcm) is crosstalk within the cable, and within the cable terminations to the repeaters. One solution is to

Table 6.5 Pair cables for analogue transmission.
(From ITU-T G.611)

Parameter	Value
Conductor diameter	1.2mm
Mutual capacity	26nF/km
Characteristic impedance @ 60kHz	178Ω
Characteristic impedance @ 120kHz	174Ω
Characteristic impedance @ 240kHz	172Ω
Attenuation @ 10°C, 120kHz	2.0dB/km

recode the standard format with a reduced high frequency content so that crosstalk effects are mitigated. This has served to significantly raise the pcm occupancy on existing cables of unit twin design.

For new digital installations, the cable pairs are split into 'go' and 'return' directions by means of an aluminium screen, i.e. a D screen or Z screen cable.

Typical characteristics of a new pair cable for 2Mbit/s systems are given in ITU-T Recommendation G.613, from which Table 6.6 is extracted. FEXT (far end crosstalk) is measured with the detector at the end of the cable remote from the disturbing source. NEXT (near end crosstalk) is measured at the same end of the cable as the disturbing source. Higher rate digital systems, 6Mbit/s and 34Mbit/s, have been used over the symmetric pair cables designed for HF carrier systems discussed in Section 6.2.4.

6.2.6 Balanced pair cables for information systems

New office blocks and industrial complexes involve the use and interconnection of many data terminal equipment and other IT products. The contending transmission technologies have been reviewed

Table 6.6 Pair cables for digital transmission.
(From ITU-T G.613)

Parameter	Value
Operational bit rate	2048 kbit/s
Construction	Unit twin
Conductor diameter	0.6mm
Nominal impedance	130Ω
Attenuation @ 20° C, 1MHz (f_0)	15.5dB/km
Attenuation @ f MHz	$15.5\left(\dfrac{f}{f_o}\right)^{1/2}$
Attenuation @ t^0	$\alpha_{20^oC}\,(1 + 0.002\,(t^{\,o} - 20))$
Sinusoidal NEXT	78 dB (nominal)
Sinusoidal FEXT	64 dB
Direct current resistance @ 20°C	63Ω/km

by (Flatman, 1988, 1989). His main points relating to pair cable are as follows.

Unscreened twisted pair (UTC) as used for telephone cabling, has been evaluated by a number of independent bodies. 10Mbit/s transmission is technically feasible over distances of up to 100m within the general office environment. Screened twisted pair (STP) can be operated at data rates of tens of Mbit/s over hundreds of metres. The advantage of twisted pair for this application is that it is sufficiently low cost, low weight and small size to plan for 'saturation' cabling at the outset and so avoid expensive upheavals as demand changes.

6.3 Coaxial pair cables

The limitations of balanced pair cables for telecommunication use at higher frequencies are the crosstalk between pairs, and the lack of predictability in transmission performance. These problems are over-come in a coaxial cable.

Considering the operation of a coaxial cable, at high frequencies skin effect causes the currents to be concentrated on the outer layer of the inner conductor and inner layer of the outer conductor. The field is then contained between the two conductors of a pair and mutual interference between cables is greatly reduced.

The effective depth of penetration of an electric current flowing over a plane surface is given by Equation 6.11.

$$\delta = \left(\frac{\rho}{\pi\mu f} \right)^{\frac{1}{2}} \tag{6.11}$$

δ is the depth at which the current at the surface I_O has reduced to I_O/e. (At a penetration of 4δ the current will only be $0.02I_O$)

ρ is the electrical resistivity in ohm cm.

μ is the magnetic permeability in Henry cm.

f is the frequency in Hz.

From this equation it will be seen that δ increases with resistivity and decreases with frequency. Table 6.7 gives the value of δ for copper.

The greater depth of penetration in a higher resistivity material, tends to offset the effect of its greater resistivity, as in Equation 6.12

$$H.F.\ resistance\ R_f = (\pi\rho\mu f)^{\frac{1}{2}} \tag{6.12}$$

Table 6.7   Effect of frequency on δ in copper

Frequency (MHz)	δ (mm) approximately
10	0.03
100	0.009
1000	0.003

Thus, although the relative d.c. resistance of aluminium to copper is 1.64, the relative HF resistivity is 1.28. Thus skin effect is the most important factor determining the functioning of a coaxial cable because it controls the resistance losses and screening efficiency at high frequencies.

6.3.1 Design for minimum loss

The primary coefficients of a coaxial cable are related to the secondary coefficients in the same way as in the balanced pair case discussed in Section 6.2.1.

$$Z_O = \left(\frac{(R + j\omega L)}{(G + j\omega C)} \right)^{1/2} \tag{6.13}$$

$$\rho = \alpha + j\beta = [(R + j\omega L)(G + j\omega C)]^{1/2} \tag{6.14}$$

At high frequencies ω is large, giving Equations 6.15 to 6.19

$$Z_O = \left(\frac{L}{C} \right)^{1/2} \tag{6.15}$$

$$\alpha = \frac{1}{2} \left(\frac{R}{Z_O} + GZ_O \right) \tag{6.16}$$

$$\beta = \omega (LC)^{1/2} \tag{6.17}$$

$$Z_O = \frac{13\delta}{\sqrt{\varepsilon}} \log \frac{D}{d} \quad ohm \tag{6.18}$$

$$\alpha = \frac{7.83 \times 10^{-2} (\rho \varepsilon f)^{1/2}}{\log_{10} \dfrac{D}{d}} \left(\frac{1}{d} + \frac{1}{D} \right)$$
$$+ 9.08 \times 10^{-5} f (\varepsilon \tan \delta)^{1/2} \tag{6.19}$$

Where ρ is conductor resistivity in ohm.cm
ε is the dielectric constant (= 1 in air)
D is the inside diameter of outer conductor

d is the diameter of the inner conductor

f is the frequency

tan δ is the power factor of the dielectric

It will be seen that only the first term of the expression for attenuation is dependent on the ratio of conductor diameters as in Equation 6.20.

$$\alpha_c = \frac{7.83 \times 10^{-2} \, (\rho\varepsilon f)^{\frac{1}{2}}}{\log_{10}q} \left(\frac{1}{d} + \frac{1}{D} \right) \tag{6.20}$$

where $q = \dfrac{D}{d}$

Then for a given attenuation α_c Equation 6.21 can be obtained.

$$D = \frac{7.83 \times 10^{-2} \, (\rho\varepsilon f)^{\frac{1}{2}}}{\alpha_c} \frac{(q+1)}{\log_{10}q} \tag{6.21}$$

By partial differentiation, D will be a minimum when Equation 6.22 holds.

$$q \, (\, \log_e q - 1 \,) \, = \, 1 \tag{6.22}$$

Therefore q = 3.59. Thus for a given diameter of coaxial cable, the attenuation is a minimum when the ratio of conductor diameters is 3.59. The characteristic impedance is given by Equation 6.23, which equals 76.7ohm for a dielectric constant of 1(air) and 50.6ohm for a dielectric constant 2.3(polyethylene).

$$Z_O + \frac{13\delta}{\sqrt{\varepsilon}} \, \log_{10} \frac{D}{d} \tag{6.23}$$

Most air spaced cables are 75ohm impedance and flexible polyethylene insulated cables of 50ohm impedance. Cables can, of course, be proportioned for different optimisations and complexities of materials (Dummer, 1961).

6.3.2 Coaxial cable construction

For general telecommunication use, coaxial cables are manufactured with a nominal impedance of 75ohms. By using a minimum amount of dielectric material, e.g. widely spaced polyethylene discs, they are effectively air spaced. The inner conductor is a solid copper wire and the outer usually formed from a folded copper tape with a longitudinal seam.

One commonly used manufacturing process incorporates the following features. The insulation between the centre and the outer conductors consists of an air filled moulding made from high density polyethylene copolymer which is continuously applied to the centre conductor of 1.2mm diameter. The insulation is formed by an on line moulding process in which two half mouldings are consolidated together around the centre conductor as shown in Figure 6.2. The effective dielectric constant of this composite insulator is 1.6. The outer conductor is formed from 0.180mm copper tape into a tube closely fitting around the centre moulding. The tape edges are corrugated in order to prevent overlap at the tube seam. The stages in the construction of the moulded shell insulation are as follows:

Stage 1. Continuously extruded strip of high density polyethylene copolymer.

Stage 2. Two half-insulations moulded into strip by embossed rollers.

Stage 3. Moulded strip divided and centre conductor introduced.

Stage 4. Two halves heat-consolidated together around centre conductor, forming single tube with integral spacing discs and two longitudinal fins.

Stage 5. Fins removed to give insulated conductors.

Two soft steel tapes of thickness 0.1mm are lapped around the outer copper, in opposite directions in such a manner as to cross each other at right angles. The inner tape has a nominal 5% gap between turns, and the outer which is wider but applied at the same lay length has a nominal 15% overlap. Two thickness of insulating tape, with colour coding are then applied. The novel method of steel tape application ensures high crosstalk immunity between coaxial pairs at

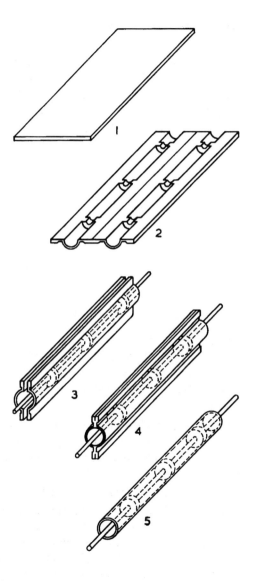

Figure 6.2 Stages in the construction of a moulded shell insulation

low frequencies and the whole manufacturing process can be controlled to ensure a consistent product.

Coaxial pairs, or tubes of the type described are laid up into cables of 4, 6, 8, 12, 18 and even 40 coaxial pairs. Balanced pairs are sometimes provided in the interstitial spaces between the coaxials for the control and supervision of the main transmission equipment.

Some operating companies combine the coaxial pairs with complete layers of balanced pairs to be used for local circuits. However, the initial saving has to be balanced against the extra complication of installation, and maintenance operations on the layer pairs. The complete cable is sheathed in polyethylene preceded by a Glover barrier of 0.2mm aluminium foil. Steel tape or wire armouring can be added if the cable is to be directly buried.

6.3.3 Coaxial cable parameters

6.3.3.1 *Impedance*

In a transmission system the source, cable and load impedances are all matched. An impedance mismatch causes a reflected backward travelling wave which serves to reduce the power of the forward travelling wave, causing an apparent increase in transmission loss. Manufactured cable lengths can be selected to be of the nominal impedance at repeater points, and of matching impedances at other joints. However, this practice is not viable on cables carrying a large number of pairs, and is not necessary if the impedance is controlled in manufacture to within 1ohm at 1MHz. The biggest source of reflections arise at the cable/repeater interface and these accumulate along a repeatered line to steep attenuation frequency rolls which are difficult to equalise. To limit these effects the ITU-T recommend that the cable and repeater impedances satisfy Equation 6.24. (G332.5).

$$N = 2A + 20\log_{10}\left[\frac{Z_E + Z_L}{Z_E - Z_L} \right] + 20\log_{10}\left[\frac{Z_L + Z_R}{Z_L - Z_R} \right] \qquad (6.24)$$

where N = 55dB minimum for a 12MHz line system and 65dB minimum for a 60MHz line system.

A = section loss at frequency f

Z_E= repeater output impedance at frequency f

Z_L = cable impedance at frequency f

Z_R = repeater input impedance at frequency f

6.3.3.2 *Impedance irregularities*

Impedance irregularities within a manufactured length of cable can arise, for example, from variations in the concentricity of the inner conductor, imperfect circularity of the inner and outer, and variation in the dielectric support.

Further disturbances can arise during factory operations prior to and during the laying up of a multipair cable, for example, roughness on an intermediate process drum.

The resultant reflections can be assessed by pulse echo testing and swept frequency reflectometer testing as a routine factory test and the product quality controlled to system requirements.

6.3.3.3 *Attenuation*

The attenuation coefficient is specified at 1MHz in dB/km. To a close approximation the loss at other frequencies is inversely proportional to the square root of the frequency. For example, the loss at 9MHz is three times the loss at 1MHz.

More exact formulae are provided for a 1.2/4.4mm and 2.6/9.5mm Coaxial Pair, in Section 6.3.4, which include a constant and an f term. Alternatively the consistency of product is such that manufacturers can provide measured data on which the equalisation of analogue systems can be based.

The attenuation variation with temperature is also important. In a buried cable, the temperature change will not exceed an annual variation of $\pm 10^o$ C, and the prevailing annual mean temperature will lie typically between $+5^o$ C and $+15^o$ C. The temperature coefficient is taken as $2x10^{-3}$ per degree centigrade at frequencies of 500kHz or more, increasing at lower frequencies to $2.8x10^{-3}$ per degree centigrade at 60kHz.

6.3.3.4 *Crosstalk*

As previously discussed, the structure of coaxial pairs is such that high values of crosstalk suppression are obtained between pairs. Poor crosstalk on an installed cable is evidence of a fault condition, such as a broken outer conductor.

The most demanding systems crosstalk requirement arises from consideration of sound programme transmission in the same carrier frequency band on adjacent coaxial pairs. In practice the limiting factor tends to be crosstalk within the carrier and repeater equipment and not within the coaxial cable system.

6.3.3.5 *Group delay*

The phase delay/frequency characteristic of a coaxial cable is very smooth and consequently the group delay distortion is low across a telephony or wideband data channel. However, both baseband and carrier TV transmission will require group delay equalisation, but the dominant contributions arise from band limiting components of the system and not from the cable.

6.3.4 Standard coaxial pair cables

Three sizes of coaxial pair have been standardised by the ITU-T for telecommunication use. These are designated by the diameter of the inner conductor and inside diameter of the outer conductor. The largest 2.6mm/9.4mm, is the oldest design and is in extensive use in many parts of the world. Initially installed for low capacity analogue systems (600/960 channel) it has been successfully exploited for high capacity analogue systems (up to 60MHz for 10,800 channel systems) and for 560Mbit/s digital systems. The 1.2mm/4.4mm design has been particularly successful as a bearer for short and medium haul routes operating at up to 18MHz (3600 channel analogue) and 140Mbit/s digital. The smallest and newest cable, 0.7mm/2.9mm, has had limited application as a 34Mbit/s digital system bearer.

The performance of these cables is summarised in Appendix 6.1 to 6.3. In each case the ITU-T recommendations make some practical distinctions between the performance demanded on a factory length and the results obtainable on a complete and installed repeater section. For the complete recommendations and explanatory notes the reader is referred to the ITU-T publication.

6.3.5 Coaxial cables for data networks

Coaxial cable offers better noise immunity and crosstalk than pair cable for cabling within a building. It can also be operated at higher bit rates and has less loss per kilometre.

A commonly used technique is to multiplex many information channels into a broadband spectrum extending up to 450MHz on a single coaxial tube. Similar transmission equipment to that used on multichannel cable TV systems can then be employed. However, system costs are high compared with alternative technologies (Flatman, 1989).

6.3.6 Coaxial cable for submarine systems

The special requirements of submarine telecommunication systems, which provide many international telephone links across oceans and shallower seas, result in a coaxial cable design rather different from the land-line product. Both cable and submerged repeater equipment have to be designed and manufactured to rigorous standards of performance and consistency coupled with high reliability and long life expectancy.

The preferred system configuration uses a single coaxial cable with the two directions of transmission transmitted in different frequency bands separated by directional filters at each repeater (Worthington, 1986).

6.3.6.1 *Deep-sea cable*

This cable is used where there is no risk of damage from fishing activity or ships' anchors. It needs no mechanical protection but must have tensile strength and be able to withstand deep water pressure. The design used by one manufacturer comprises a central strength member of high tensile steel wires which can tolerate the loads of 10 tons or more arising when a repeatered cable is recovered in a rough sea. This is covered in a copper tape welded longitudinally forming the inner conductor. The dielectric is solid polyethylene which, as well as having excellent electrical properties, is virtually incompressible. The outer conductor is a single metal tape closed in a longitudinal overlap.

This can be of copper, but aluminium is often used. The attenuation is then increased by 6%, as is the number of repeaters, but the overall system costs will be lower. Finally, an overall polythene sheath is extruded to apply compression to the outer conductor and provide some abrasion resistance. The dimensions of such a cable are considerably greater than standard land-line cable. Thus the inner conductor has an outside diameter of approximately 9.3mm, the dielectric a diameter of 37.3mm, the outer tape a thickness of 0.46mm and the overall diameter over the sheath is 44.5mm.

6.3.6.2 *Shallow-water cable*

In shallower seas, i.e. less than about 1000m, the risk of damage from ships' anchors and fishing activity, particularly trawling, can be high. The solution is to add one or two layers of mild steel armouring wires to the lightweight deep-sea cable, packed in a polypropylene bedding.

Sometimes in vulnerable situations a trench is ploughed on the sea floor as the cable is laid and the cable buried for extra protection.

6.3.6.3 *Cable performance*

This cable design conforms with the outer to inner conductor diameter ratio of 3.6 to 1 for minimum loss as derived in Section 6.3.1 above. As the dielectric is solid polythene, the characteristic impedance of the cable is 50ohm. The pressure coefficient of the cable is remarkably low at about 0.5% per 2km of sea depth.

This cable has been used for 1840 circuit systems operating up to 13.7MHz and for 5520 circuit systems operating up to 44.3MHz. The cable loss is given in Table 6.8. (The standard unit of length in

Table 6.8   Loss of 37.3mm submarine coaxial cable at 10°C

Parameter	Value			
Frequency	1	5	15	45
Loss (dB/nm)	1.69	3.84	6.84	12.54
Loss (dB/km)	0.91	2.07	3.69	6.76

submarine system work is the nautical mile (n.m.) of 6087 feet (1.855km).) Laboratory studies demonstrated that systems of much higher frequencies were feasible, but this work has been dropped in favour of the development of optical fibre systems.

6.4 Line plant for copper cables

Cables can be installed underground or overhead. For overhead installation the cable can be lashed tightly against a previously installed messenger wire. Alternatively, the messenger wire, or catenary, can be built into the cable sheath to give a figure-of-eight section and clamped to each supporting pole. Cable joint housings are usually mounted part way down the pole.

Two techniques are used for underground installation, by duct and by direct burial. Ducts have the advantage that additional cables can be added and obsolete cables removed relatively easily. Ducts are the normal solution in urban areas and are preferred by some operating companies (e.g. BT) for virtually all applications. For many years lead was the preferred cable sheath as it is totally impervious and waterproof joints can be made.

Plastic, particularly polyethylene offers lower cost, greater toughness and lighter weight, but its widescale adoption had to wait until sheath jointing by either welding or epoxy adhesives were developed. The polyethylene sheath is applied over and bonded to an aluminium/polyethylene laminate moisture barrier. This aluminium foil also provides the equivalent electrical screen to a lead sheath.

For direct burial the cable is armoured with layers of steel wire or tape appropriate to the installation conditions. In open country cable ploughs are employed which open the trench, lay the cable and backfill in one operation.

Two techniques are used to resist water penetration through a damaged sheath. The cable core can be saturated with petroleum jelly or a similar compound. The disadvantage is the difficulty of cleaning all the cable elements when joining and it is not appropriate to air cored coaxial cables. The second technique is by air pressure, either continuous flow or static. Static pressure systems rely on total air integrity of the cable. Regularly spaced pressure sensitive switches e.g. at each joint, close across a monitor control pair when pressure drops. Resistance measurements from the terminal stations identify the location. Continuous flow systems tolerate some leakage from the cable and damage is sensed by an increase in flow rate. Cables to be used with either of these systems must have a sufficiently low pneumatic resistance.

A repeatered cable line includes accommodation for equipment e.g. PCM regenerators at 1.8km on pair cable; coaxial line repeaters spaced in the range 1.5km to 9km on large core cable and 2.0 to 4.0km on small core cable. These are robust steel or cast iron housings with a removable lid which when sealed will withstand several metres of water pressure. Several housings may be used to support one large cable which is separated out into individual pairs at a segregating joint.

The individual tail cables have to be air tight and to retain the electrical characteristics of the main cable. For convenience of equipment installation and maintenance, the housings are best mounted to one side of the main cable in a dedicated shallow manhole or footway box. Water tightness of the repeater housings is assured by air pressurisation which may be integrated into the main cable system.

6.5 Optical fibres

Telecommunication transmission systems operating over light guiding optical fibre waveguides were first proposed in 1966 by Hockham and Kao working at the STC Technology Ltd. laboratories in Harlow,

Essex (now part of BNR Europe Ltd). One of the earliest and most ambitious field trials of an optical fibre system was the 140Mbit/s system designed, manufactured and installed by STC plc between Hitchin and Stevenage in 1977. Since then the technology has progressed to the extent that optical fibres have virtually ousted metallic cables for new telecommunication work except for the local area network and short distance applications.

Optical fibres, or light guides, operate by total internal reflection at a core cladding boundary where the refractive index of the cladding is less than that in the core. A multimode fibre supports many hundreds of modes which can be conceptually considered as light rays. Each ray is totally reflected at the core/cladding boundary and progresses along the fibre by multiple reflections back into the core. Axially injected rays experience fewer reflections than rays injected at an angle to the axis.

Before considering the properties of optical fibres, it is useful to have some appreciation of how fibres are made.

Fibre manufacture requires a high purity glass which is generally achieved by fabrication from compounds in a gaseous form. A typical process is by chemical vapour deposition (CVD) the principles of which are as follows: silicon tetrachloride, oxygen and germanium chloride gases are fed into a silica tube which is rotating in a lathe. A hot burner traverses along the tube causing a chemical reaction at that point and glass to be deposited on the inside wall of the tube. At each traverse of the burner the percentage of germanium dopant can be modified and the refractive index of the glass changed. Other dopants are phosphorous and fluorine. A hundred or more passes are typically involved and the whole process is computer controlled. Finally the burner temperature is raised and the support tube collapses into a solid rod or preform.

It will be apparent that an advantage of this technique is that as well as pure starting materials, the complicated deposition and collapse process take place in a closed environment, minimising the problems of contamination by handling or from the atmosphere.

In the next operation, the preform is fitted in a pulling tower and its end heated to a temperature at which a fibre filament can be drawn off, at about $2200^{o}C$. At the bottom of the tower is the take up drum the speed of which is controlled by a diameter monitor just below the drawing furnace. Following the diameter monitor, the fibre is drawn through a primary coating vessel, curing oven, and then finally the fibre is wound on the take up drum. It will be noted that the primary protective coating is applied virtually as the fibre is drawn.

6.5.1 Features of multimode fibres

Standard multimode fibre for telecommunication application has a core diameter of 50μm and a cladding surface diameter of 125μm.

The numerical aperture (N.A.) of a fibre is a measure of the angle of acceptance of the light rays presented to it as shown in Figure 6.3 which illustrates numerical aperture. Rays at too steep an angle will not be totally reflected and not be guided.

If θ_c is the critical angle for total reflection, and θ_a is the maximum incident angle, then the numeric aperture is given by Equation 6.25.

$$N.A. = \sin \theta_a = (\eta_{co}^{\ 2} - n_{ce}^{\ 2})^{\frac{1}{2}} \tag{6.26}$$

Where η_{co} is the maximum refractive index of the core and n_{ce} is the minimum refractive index at the cladding boundary. The larger

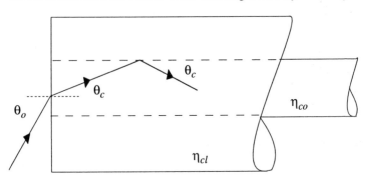

Figure 6.3   Illustration of numerical aperture

the N.A., the more modes the fibre will support but the baseband response may be limited. Fibres for telecommunication purposes have an N.A. of typically 0.2.

The simplest form of multimode fibre is a step-index design in which there is a well defined step in the refractive index between the core glass and the cladding. In such a fibre the light rays (highest order modes) experiencing most reflections travel a longer path down the fibre than rays experiencing fewer reflections (lower order modes). Consequently the arrival time of the higher order modes at the end of the fibre is later. A transmitted pulse at the output end of a fibre will be longer than that inputted. To an analogue signal this effect shows as a bandwidth limitation.

The most commonly encountered form of multimode fibre is the graded index type in which the refractive index decreases with offset from the centre to an approximately parabolic law. The difference between the refractive index at the axis and the boundary of the core is about 1%. This has two effects. Firstly, the fibre becomes self focusing and secondly, transit speed in the outer, lower refractive index regions of the fibre is faster than in the higher refractive index region at the core. That is, there is an equalising effect on the transit times of the higher order modes and the pulse spreading effect is reduced by a factor of 100.

For many fibres the pulse spreading which results from these multipath differential propagation times is of a Gaussian shape. Then, 1ns of pulse distortion measured at half amplitude, corresponds to a drop of 6dB measured in the amplitude/frequency response at 440MHz.

In an installed cable of jointed fibre lengths, mode mixing takes place at the joints and at small irregularities within the fibre lengths and this has the effect of reducing the effects of modal dispersion. The summation of modal dispersion in a real route is given by dL^{c} where d is the dispersion per kilometre, L is the length in kilometres and c is the concatenation index. The concatenation index for most fibres lies between 0.5 and 1.0 and is usually taken as 0.7.

Commercially the modal dispersion of multimode fibres is usually expressed in terms of bandwidth per kilometre. A typical grading would be from 400MHz to 1600MHz in 400MHz steps.

The refractive index of glass is a function of the light wavelength. As the propagation rate is a function of the refractive index it follows that different wave lengths have different propagation speeds in the fibre.

Thus pulse dispersion arises from variation of the refractive index across the band of wavelengths emitted by the light sources used in the equipment, i.e. lasers and LEDs.

A typical value for graded index fibres is 120ps/km per nm at a centre wavelength of 850nm, that is a pulse emitted from a 10nm width source would be subjected to 1.2ns widening in a 10km length. With increasing wavelength, material dispersion in silica fibres falls and becomes negative above about 1270nm. As attenuation also falls with wavelength, 1300nm is an attractive operating window. The value of material dispersion at 1300nm is usually taken as 6ps/km nm.

Material dispersion in the next transmission window of 1550nm where attenuation is still lower, is around 17ps/km nm. By increasing doping concentrations, the material dispersion zero can be shifted to 1550nm, but at the expense of some slight increase in attenuation at 1550nm (See Section 6.4.2. on single mode fibres).

In systems planning calculations, the effects of both material and modal dispersion can be considered to be of Gaussian shapes and added by taking the square root of the sum of the squares.

Low attenuation coupled with wide bandwidth, is probably the most remarkable property of optical fibres and accounts for their outstanding success. Values of around 0.5dB/km at 1300nm are commonly achieved on cables for general telecommunication applications, and even lower values at 1550nm for submarine telecommunication systems.

There are two basic loss mechanisms in optical fibres. These are absorption and scattering:

1.   Absorption loss. Pure silica is transparent in the wavelength range from approximately 200nm to 10000nm, with a maximum transparency at a wavelength of 1550nm. Small traces of metallic impurities in the silica can increase the loss of the fibre considerably and metallic ion (fe, Cu, Ni etc.) concentrations must be limited to less than 0.01ppm. A common im-

purity is water in the form of hydroxyl ions (OH) which can produce rather wide absorption bands of overtones of the 2800nm fundamental which can typically be seen around 1400nm, 1250nm, 970nm and 750nm. In modern fibres the water content can be reduced to less than 1 part in 10^8.

2.   Scattering loss. Rayleigh scattering of the light occurs within the molecules of glass material itself. This loss is independent of the light intensity but varies inversely with the fourth power of wavelength. Hence the advantage of operating at longer wavelengths. These loss mechanisms result in three operating transmission windows in the loss spectrum of silica optical fibre in the region of 800nm, 1300nm and 1500nm. Historically 850nm was the first to be commercially exploited and 1550nm is the most recent. However, interest continues in the use of the 850nm window for short distance applications because the electro-optic devices are considerably cheaper. Most commercial activity is directed at 1300nm with 1550nm limited to applications where very long spans are warranted, such as submerged cable systems.

6.5.2 Single mode fibre

The bandwidth limitations of multimode fibres arise from the variation in multipath propagation times. The best graded index fibre offers a bandwidth of about 1.5GHz/km and is expensive and difficult to produce. (On a 30km route this would reduce to something like 140MHz which is inadequate for high bit rate systems.)

This major limitation is overcome in single mode fibre designs. The number of modes propagated by a fibre is given approximately by Equation 6.26.

$$N = \frac{2\pi^2 a^2}{\lambda^2} \left(\eta_{co}^2 - \eta_{cl}^2 \right) \tag{6.26}$$

where a is the core radius

λ is the wavelength

η_{cl} is the refractive index of cladding

η_{co} is the refractive index of the core.

As the fibre diameter is reduced, the number of modes which can be propagated falls, and in the extreme only a single mode is transmitted. In order to obtain a usable core size the core cladding index difference is also reduced from the order of 1% to 0.1%.

In a single mode fibre the core size is about one tenth of that of a graded index fibre (5 micron) and obviously the practical problems of injecting light into the fibre, jointing fibres and connector design are more difficult. However, solutions are now well established and single mode fibres have become the standard for virtually all new telecommunication work.

In a single mode fibre, part of the power is carried in the fibre cladding, and is still guided. The concept of mode field diameter is therefore more useful than core diameter and can be defined as the width between the points across the fibre where the optical field amplitude measured is $\frac{1}{e}$ of the maximum value.

The overall diameter over the cladding of the fibre as drawn is 120 micron, the same as for standard multimode fibre. The cladding is thus very much larger than the core and is of optically low loss. One consequence of this is that care has to be taken to exclude cladding light when making measurements, for example by taking the fibre through a bath of liquid of higher refractive index than the cladding. Some fibre protective coatings are also designed to act as cladding mode strippers.

A single mode guide will only be single mode above a certain wavelength and below this value second order modes will also be propagated. The cut-off wavelength is usually in the range 1100-1280nm.

Reference has already been made to dispersion shifted fibres in the earlier section on material dispersion. This involves increasing the dopant concentrations to shift the naturally occurring zero value at 1270nm in silica fibre to the lower loss window of 1550nm. However, this also serves to increase the loss at 1550nm. The effect can be mitigated by changing the refractive index profile of the fibre from a rectangular to a triangular shape.

Single mode fibre provides a low loss and high bandwidth transmission bearer which is economic over a wide range of digital bit rates using either laser or LED sources, and thus provides an operating administration with a "future-proof" investment.

The low values of pulse dispersion make system planning a relatively straightforward task up to about 600Mbit/s. At higher bit rates additional allowances may have to be made for very short term laser wavelength changes during the 'on' period which are converted by the small but finite dispersion of the fibre to an additional noise source.

6.6 Optical fibre cables

There are two primary considerations determining the design of an optical fibre cable:

1. To minimise optical attenuation increments associated with the manufacture and use of the cable.
2. To maintain the physical integrity of the fibre during the cabling process and its subsequent installation and service environment.

Macrobending loss arises due to a partial loss of guidance in the fibre when it is bent into a curve having a radius less than a few tens of millimetres. When the fibre is so bent, the evanescent wave in the cladding needs to travel above the velocity of light in the medium, which it cannot do, and so guiding is lost and energy is radiated. At wavelengths approaching the mode cut-off wavelength, more energy is carried in the evanescent field in the cladding and bending loses more easily occur. In multimode fibres, there are always some modes close to cut-off and less tightly bound than the lower order modes. Minor perturbations in the fibre geometry causes coupling between modes and hence replacement of those vulnerable to loss.

Microbending loss occurs when a fibre is in contact with a rough surface that imposes many very small bends on the fibre. Bending loss can be reduced by designing the fibre to have a small mode field diameter, but this conflicts with the need for a large spot size to reduce splicing and connector difficulties.

Standard fibre has a mode field diameter between nine and ten micrometres.

Glass fibre, unlike steel or copper wire, does not have a well defined or controllable breaking strain because failure is initiated in flaws that are usually at the surface, but may also be within the fibre. These flaws cause local intensification of any applied stress.

For a typical telecommunication fibre, a flaw only one micrometre deep will cause failure at about 1% strain. Flaws arise from a number of sources during fabrication, for instance: imperfections in the substrate tube, scratches introduced by handling, discontinuities during deposition, stresses introduced during collapse or pulling, impurities implanted from the furnace during pulling and foreign matter trapped during coating.

Below the critical strain at which a flaw will cause immediate fracture, the flaw will increase in size until fracture occurs. It can be shown (Allard, 1990) that the time to failure is inversely proportional to the nth power of the applied stress or strain, where n lies in the range of 14 to 25, the higher value being in dry conditions and the lower value at 100% humidity.

Fibres with the largest flaws can be eliminated by applying a proof test, typically of 0.6% strain, as an online manufacturing process, and ensuring by cable design that the stress on the fibre at all stages in its life remains low.

A fibre buffer or primary coating serves the dual purpose of protecting the fibre from rough surfaces which would introduce microbending losses, and, of preserving the pristine strength of the fibre. Hence, the coating is applied on the pulling tower within a second or two of the fibre being drawn in such a way that the fibre is not scratched or marked.

Thermally cured polydimethyl silicone (e.g. Sylgard) provides a marked increase in a usable fibre strength and very small temperature dependence. More recently acrylate resin coatings have been widely adopted as these offer advantages in fibre drawing speeds and in more versatile buffering techniques. A typical system employs a low modulus inner coating with a high modulus outer coating, giving a package which provides a tough smooth exterior capable of with-standing subsequent handling and abrasion, while the soft inner protects against microbending losses.

The coated fibre requires further protection before it can be incor-porated into a cable. A common solution is to extrude a loose tube over the fibre, and if the fibre is made to lie in a long helix within the tube by overfeeding, strain relief of 0.2 to 0.4% can readily be obtained. However, there is clearly some uncertainty on the positional stability of the fibre, and contacts between the fibre and inside wall of the tube (which is relatively rough) can give rise to microbending loss.

Long term movement and water blocking problems can be over-come by introducing a gel into the tube. Thixotropic gel compounds have been especially developed for optical fibre cables (Bury, 1985) with a wide temperature range and a fine granular structure. The gel is normally highly viscous and supportive of the fibre, but changes to low viscosity under stress and allows rapid movement of the fibre during cable elongation of compression. With this arrangement good protection of a fibre with a single hard acrylic coating is obtained.

A common alternative to loose tube encapsulation is tight buffer-ing. In this the coated fibre has extruded an additional jacket of polymer such as nylon 12 to an overall diameter of between 0.5 and 1.0mm. This can be applied in such a manner to apply a longitudinal compression to the fibre and obtain a strain relief of 0.2% or more.

Multifibre encapsulation in a gel filled loose tube for high fibre count cables is also attractive e.g. up to 12 fibres can be inserted in a tube of 2mm inside diameter and gel filled.

6.7 Cable design

With the essential building block of well buffered and encapsulated fibres, optical cables can be manufactured in a similar way to conven-tional cables.

A design for installation in long underground ducts is shown in Figure 6.4. Tight jacketed buffered fibres are laid helically around a strength member of stranded steel wires. The cable is pulled through the duct via a mechanical fuse which ruptures if the intended stress is

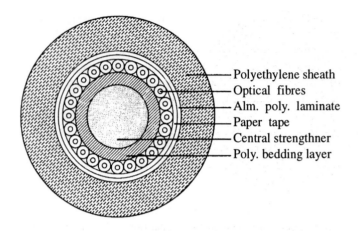

Polyethylene sheath
Optical fibres
Alm. poly. laminate
Paper tape
Central strengthner
Poly. bedding layer

Figure 6.4 Tightly buffered optical fibre trunk cable

exceeded, which is well before stress on the fibres exceeds 0.2% to 0.3%. Attenuation increments during the cabling process are very low. The same general design is also used with fibres encapsulated in loose tubes and in gel-filled loose tubes (Figure 6.5). For applications where non-metallic cables are required - alongside electrified railways or in regions of high lightning activity, the aluminium plastic laminate is omitted and dielectric strength members (e.g. Kevlar) replace the steel core.

The slotted core construction (Figure 6.6) is effective for high fibre count cables. Figure 6.7 shows a more recent design with several fibre bundles laid in a large protective soft gel-filled tube. This cable is for direct burial and is shown completed with outer sheath's lightning and rodent protection.

6.7.1 Self supporting cable

The advantages of interference free operation have encouraged the installation of optical fibres on high voltage power lines. The high voltage power transmission and distribution network provides an alternative communication route for use by common carriers. In an

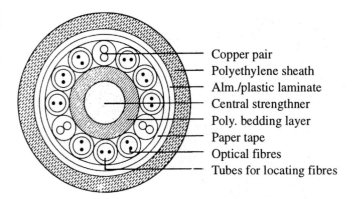

Figure 6.5   Optical fibre loose tube cable

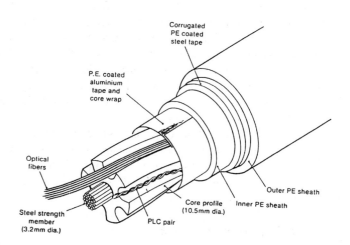

Figure 6.6   Slotted core cable with rodent protection

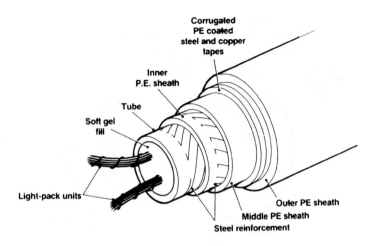

Figure 6.7   Alternative cable design with rodent protection

earlier development, optical fibres have been incorporated into over-head line conductors, but application is limited to new or recon-structed routes.

The cable shown in Figure 6.8 (Rowland, 1987) has been de-veloped by one manufacturer for economic retrospective installation on live power transmission lines operating at voltages up to 150kV.

The strength member is a pultruded glass reinforced plastic (grp) rod at the top of which is a small rectangular slot. Within this gel filled slot are two twelve-fibre ribbons covered with a cap. The cable is sheathed in a track-resistant material to an overall diameter of only 13mm. In temperate climates, subject to simultaneous ice and wind loading the maximum span between towers is 550m and 720m over river crossings. Without ice loading, spans can be significantly larger.

The optical loss of the fibres is designed to remain within 0.5dB/km at 1300nm during all conditions of service. This includes temperature excursions from -40°C to +70°C and a 25 year service life. The cable is supplied in 4km lengths and supported on the bottom crossarms. The installation fittings and procedures are compatible with safe working practices on live routes.

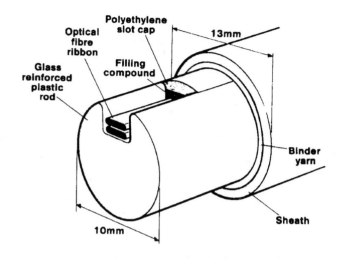

Figure 6.8 Self-supporting optical fibre cable

Ongoing research is expected to extend the area of application of this self supporting all dielectric communication cable to include 400kV power lines.

6.7.2 Submarine optical cables

A second example of an optical fibre cable for special application is one for submarine system use. These are now in service in both shallow waters and in trans-oceanic links. Consideration has to be given to the high strains during installation and in any recovery and repair operations. In shallow waters additional protection must be provided against damage by anchors and trawling operations.

An example of cable manufactured by one manufacturer is shown in Figure 6.9 and the optical characteristics are given in Appendix 6.4 (Mazda, 1989). An interesting feature of this cable is the copper tube around the fibres which is the system power feeding conductor and also a hydrogen barrier.

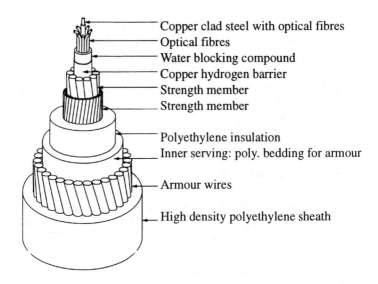

Copper clad steel with optical fibres
Optical fibres
Water blocking compound
Copper hydrogen barrier
Strength member
Strength member

Polyethylene insulation
Inner serving: poly. bedding for armour

Armour wires

High density polyethylene sheath

Figure 6.9 Submarine optical fibre cable

Hydrogen causes adverse changes to the attenuation of optical fibre (Barnes, 1985). This is of no consequence on land line cable, but of much greater significance in a deep sea cable because of the high pressure. Loss mechanisms arise from interstitially dissolved hydrogen, hydroxyl formation, and the formation of defect centres. The hydrogen can arise by galvanic corrosion, metallic outgassing, polymer degradation, magnetic hydrodynamic effects and other causes. By careful choice of fibre and cable materials, and by appropriate design the hydrogen effect can be limited to a tolerable level (Worthington, 1986).

6.7.3 Optical cable for information systems

Much attention has been devoted to using optical fibre technology in a 10Mbit/s ethernet. Another current project is FDDI-II (fibre distributed data interface) which is a 100Mbit/s dual optical fibre ring (Flatman, 1989).

Clearly the wideband transmission characteristics of optical fibre, and its immunity to interference make it an ideal medium for this work. The inhibiting factors are the relatively high cost of optical devices and connectorisation in the context of short distance, high density links. An elegant solution to this problem has been reported by workers at STC Technology Limited (Cannell, 1988). Access to a single mode fibre is accomplished via a non-intrusive tap and a 10Mbit/s ring with a capacity of 100 nodes is described.

6.8 Line plant for optical cables

Generally it has proved feasible to adapt conventional cable hardware to the requirements of optical fibre. The preferred fibre jointing technique is by fusion of the two fibre ends, either in a gas flame or more commonly by an electric arc. Proprietary machines are available in which the fibre coatings are stripped, the fibre ends cleaved, and fusion applied. Surface tension forces centre the mating fibres with respect to the cladding surface diameters. Finally, the surface coatings are reinstated and the splice given mechanical support. Machines with various levels of automation are available and some provide light injection and detection (LID) so that splice loss can be minimised during the operation. It is common practice to check splice losses by OTDR (optical time domain reflectometer) progressively as work proceeds along a route.

The completed splice, with its reinforcement splint, must be carefully stowed within a splice housing and organiser. It is normal to allow surplus fibre for a splice remake and this too must be stowed within the constraints of the minimum bending radius specified. The complete splice housing is finally rendered integral with the cables by polyethylene injection welding or heat shrink techniques.

Intermediate repeaters on long optical cable routes can readily be accommodated in the same design of underground housings used for metallic cable systems. It is only necessary to design an appropriate pressure tight cable termination and tail cable which is spliced to the main cable. With repeater spans up to 30km it is feasible to power feed these repeaters over 0.9mm copper conductors in the cable. Although still practical at longer spacings, the voltage drop in the cable becomes increasingly significant and local power sources, such as solar power can be considered as an alternative.

6.9 ITU-T standardised optical fibre

Probably the most important contribution of the ITU-T to the rapid development of fibre optic technology has been the achievement of internationally accepted definitions and reference test methods (RTM).

The recommended fibre characteristics in G.651 and G.652 are generally sufficient to ensure compatibility between conforming cables and manufacturers' optical repeater equipment. Key parameters from these recommendations are given in Appendix 6.5 and 6.6.

6.10 Future developments

The current standard single mode telecommunication fibre provides remarkably low attenuations of 0.5dB/km in the 1300nm window and, optionally, 0.2dB/km at 1500nm. These values, in conjunction with a variety of transmitters and receivers of different powers, sensitivities and costs, enable system designers to produce economic and technically attractive solutions to most telecommunication needs. This fibre is the culmination of many years development in laboratories throughout the world, and it is unlikely that it will be superseded as a general transmission bearer for some years. Optical device development will certainly continue and new system designs offering advantages in transmission bit rates and costs will continue to emerge. Cable designs incorporating standard fibre will also continue to evolve in response to new opportunities and special needs.

There are however two, perhaps three, market sectors of special interest where a new fibre design might be justified. The first is in submarine cable systems where the benefit of lower loss means fewer underwater repeaters and potentially higher reliability. Some research into new glass compounds offering very low losses at long wavelengths (10 μm) has been reported. The second market sector is the proposal for optical links into telephone subscribers premises. This application would probably benefit from larger core fibres facilitating the design of the necessarily low cost optical devices and connectors. A third possible area is application in local area networks which is in some respects similar to the problem of subscriber connection. However, current sentiment seems to be in favour of standard fibre for this application.

It is evident that any alternative fibre design in terms of operating wavelength or core size, must be supported by a corresponding range of transmitters, receivers, connectors and test equipment - and that the resulting system has to be cost competitive with standard alternatives. Technological prediction is a dangerous game, but it would seem that today's standard fibre has an assured future by virtue of its outstanding performance and the massive investment already committed by operators in their cable networks.

6.11 Acknowledgements

The writer wishes to thank Mr. M.M. Ramsay and Dr. John Lees of BNR Europe (formerly STC Technology) for their generous assistance, and the directors of NT Europe (formerly STC) for permission to publish this paper.

6.12 References

Allard, R.C. (1990) *Fiber Optics Handbook for Scientists & Engineers*. McGraw-Hill, Section 2.1.1.

Barnes, S.R. et. al. (1985) The Effect of Hydrogen on Submarine Optical Cables. *Electrical Communication* **59** (4).

Biederstedt, L. (1994) The advantages of shielded cabling systems, *Telecommunications*, June.

Bonicel, J.P. (1994) Optical ground wires and self-supporting all-dielectric fibre optic cables, *Electrical Communications*, 1st Quarter.

Bury, J.R. and Joiner, D.A. (1985) Versatile High Performance Filling Compounds for Telecoms Cable Applications. *In Proceedings of the International Wire and Cable Symposium.*

Cannell G.J. et. al. (1988) Access Methods for Non-Intrusive Optical Fibre Networks. *In Proceedings of the International Wire and Cable Symposium.*

CCITT (1989) *Blue Book.* Vol 111, ITU Geneva.

Dummer, G.W.A. and Blackband, W.T. (1961) *Wires and R.F. Cables.* Pitman.

Flatman, A.V. (1988) Universal Communications Cabling - A Building Utility. *ICL Technical Journal* pp. 117-136.

Flatman, A.V. (1989) Open Systems Cabling as a Building Utility. *Electronics and Communication Engineering Journal (IEE)* **1** pp. 152- 158.

Gijsbrechts, H. (1993) International cabling standards, *Telecommunications,* December.

Mazda, F.F. (1981) *Electronics Engineers Reference Book 5th Edition,* Butterworth Scientific, Section 56.18.

Mazda, F.F. (1989) *Electronics Engineers Reference Book 6th Edition,* Butterworth Scientific, Section 55.18.

Reinaudo, Ch. (1994) Undersea cables: a state-of-the-art technology, *Electrical Communications,* 1st Quarter.

Rowland, S. et. al. (1987) The Development of a Metal-Free Self-Supporting Optical Cable for Use on Long Span, High Voltage Overhead Power Lines. *In Proceedings of the International Wire and Cable Symposium.*

Schelkunoff, S.A. The Electromagnetic Theory of Coaxial Transmission Lines and Cylindrical Shields. *Bell Syst. Tech. J.* **13**.

Worthington, P. et. al. (1986) The Design and Manufacture of Submarine Optical Cables in the U.K. *In Proceedings of the Sub-Optic Conference, Paris.*

6.13 Appendix 6.1

2.6/9.5mm coaxial pair (ITU-T G.623) characteristics

(a) Factory Length

Impedance
Characteristic impedance is given by Equation A6.1.

$$Z = 74.4 \left[1 + \frac{0.0123}{\sqrt{f}} \ (1-j) \right] \quad ohm \qquad \text{(A6.1)}$$

Where f is the frequency in MHz, and the tolerance on impedance is ±1ohm.

Attenuation
The nominal attenuation is 18.00 ±0.3dB/km at 60MHz at 10°C.
Nominal at 1MHz is 2.32dB/km.
Above 1MHz it is $0.01 + 2.3\sqrt{f} + 0.00ff$ dB/km.

Construction
Inner diameter is 2.6mm solid wire.
Outer conductor is 0.25mm copper tape.
Internal diameter is 9.5mm over insulation.

Impedance regularity
A cable suitable for 60MHz analogue or 560Mbit/s digital systems must exhibit reflections of at least 54dB down on the inputted signal, measured with a 2ns pulse.

(b) Installed cable

Dielectric Strength
The pair shall measure not less than 5000Mohm-km at 500V dc, after electrification for one minute at 2000V dc. (Transmission systems frequently incorporate high voltage power feeding of dependent repeaters).

6.14 Appendix 6.2

1.2/4.4mm coaxial pair (ITU-T G.622) characteristics

(a) Factory Length

Impedance
The characteristics Impedance is 75ohm ±1ohm at 1MHz.

Typically, then Z at 60kHz is 79.8ohm; at 500kHz it is 75.8ohm; at 4.5MHz it is 74 ohm; at 18MHz it is 73.5 ohm.

Attenuation
Nominal attenuation at 1MHz is 5.3dB/km ±0.2dB/km.
Attenuation above 2MHz is $0.07 + 5.15\sqrt{f} + 0.005f$.

The lay up factor causes the attenuation of multipair cables to be about 0.5% greater per sheath km.

Construction
Inner conductor is 1.2mm diameter copper wire.
Outer conductor is 0.15mm or 0.18 copper tape.
Internal diameter is 4.4mm over insulation.

Impedance Regularity
The most demanding requirement is for 140Mbit/s digital operation. Reflections from a 10ns pulse shall be at least 49dB down.

Dielectric Strength
The pair shall measure not less than 5000Mohm-km at 500V d.c. after electrification for one minute at 1000V d.c. If in normal use the outers are not earthed, outers to sheath shall withstand 2000V d.c. (Isolated outers are one technique for reducing induced currents from railway traction and lightning strokes).

6.15 Appendix 6.3

0.7/2.9mm coaxial pair (G.621) characteristics

(a) Factory Length

Impedance
The characteristics impedance at 1MHz is 75 ±2.5 ohm.
Typically, then Z at 0.2MHz is 77.7ohm; Z at 5MHz it is 73.4 ohm; Z at 20MHz it is 72.8 ohm.

Attenuation

Nominal attenuation at 1MHz is 8.9dB/km at 10°C.

Typical attenuation at 0.2MHz is 4.5dB/km; at 0.5MHz it is 6.5dB/km; at 5MHz it is 19.8dB/km; at 20MHz it is 39.6dB/km.

Mechanical Construction

Inner is solid copper wire 0.7mm diameter.
Outer is 0.1mm thick copper tape, laid lengthwise.
Screen is 0.1mm thick steel tape, laid lengthwise.
Insulation has an external diameter of 2.9mm.

Impedance Regularity

The reflection from a 100ns pulse shall be at least 39dB down.

(b) Installed Cable

Crosstalk

The near-end crosstalk attenuation between coaxial pairs, measured in the frequency band 0.5 to 20MHz on 2km or 4km sections, should be greater than 130dB.

Dielectric

The pair shall measure not less than 5000Mohm-km at 500C, after electrification for one minute at 1000V d.c. Outer to screen shall withstand 2000V d.c. for 1 minute.

6.16 Appendix 6.4

Submarine optical fibre characteristics

Core diameter = 8.4μm (germanium doped).
Refractive index difference = 0.004
Fibre diameter = 125μm
Cut-off wavelength = 1.15 - 1.28μm
Wavelength of zero dispersion = 1.315μm
Attenuation at 1.310μm = 0.34dB/km
Attenuation at 1.55μm = 0.19dB/km

Dispersion at 1.55µm = 17ps(km.nm)
Mode field diameter = 9.8µm

Alternative Dispersion Shifted Fibre

Core diameter = 6.9µm (at base of triangular profile)
Refractive index difference = 0.011
Cut-off wavelength = 0.84µm
Wavelength of zero dispersion = 1.53µm
Attenuation at 1.55µm slightly greater than 0.19dB/km

6.17 Appendix 6.5

Multimode graded index optical fibres (ITU-T G.651) key parameters

Core diameter = 50µm, deviation less than ±6%
Cladding surface diameter = 125µm, deviation less than ±2.4%
Concentricity error less than 6%
Core non circularity less than 6%
Cladding non circularity less than 2%
Attenuation at 850nm generally less than 4dB/km
Attenuation at 1300nm generally less than 2dB/km
Modal bandwidth 850nm generally greater than 200MHz.km
Modal bandwidth 1300nm generally greater than 1000 MHz.km
Chromatic dispersion 850nm generally less than 120ps/km.nm
Chromatic dispersion 1300nm generally less than 6ps/km.nm

NOTE

1. The optical performance quoted is for guidance only. Fibres offering 2 to 2.5dB/km at 850nm and 0.5 to 0.8dB/km are available. Similarly, bandwidths of greater than 1000MHz.km at 850nm and 2000MHz.km at 1300nm have been achieved.

2. The ITU-T recommend specific dimensional tolerance to ensure that fibres can be interconnected with an acceptably low loss. Provisional results indicate that acceptable splice loss

and adequate strength can be achieved when splicing different high- silica fibres.

6.18 Appendix 6.6

Singlemode optical fibre cable (ITU-T G.652) key parameters

Cladding diameter = 125μm, deviation less than ±2.4%

Mode field diameter = In range 9mm to 10mm at 1300nm

Mode field concentricity error less than 1.0μm (up to 3.0μm may be appropriate for some jointing techniques and joint loss requirements).

Cladding non-circularity, less than 2%

Cut off wavelength (fibre) in range 1100 - 1280nm

Cut off wavelength (cable) less than 1270nm

Loss at 1300nm less than 1dB/km (typical)

Dispersion (1270 - 1340nm) 6ps/km.nm max

Dispersion (1550nm) 20ps/km.nm max

Bend loss at 1550nm. The loss at 1550nm shall not increase by more than 1dB when measured on 100 turns of fibre loose wound on a 75mm diameter mandrel.

7. Fibre optic communications

7.1 Principles of light transmission

7.1.1 Basics of optical fibre transmission

The propagation of light is governed by Maxwell's equations but a good insight of the propagation in dielectric material can be gained through Snell's law on refraction and reflection.

Consider two media I and II with refractive indices, n_1 and n_2 respectively. Then according to Snell's law the incident and refracted light rays satisfy Equation 7.1, and in general there is also a reflected ray which propagates in medium I.

$$n_1 \sin \varphi_1 = n_2 \sin \varphi_2 \tag{7.1}$$

Consider now the case where $n_1 < n_2$, as in Figure 7.1. Then as the angle φ_1 approaches a critical value φ_{cr} the refracted ray approaches the extreme angle of $\pi/2$ and at this point transmission of light into medium II ceases completely. This critical angle is given by Equation 7.2.

$$\sin \varphi_{cr} = \frac{n_2}{n_1} \tag{7.2}$$

When the incident angle $\varphi_1 > \varphi_{cr}$ no light propagates into medium II and all the incident light is reflected back into medium I. This basic behaviour at the interface of media I and II with $n_1 > n_2$ can be used to guide the propagation of light.

The most practical structure for the propagation of light over long distances has been the optical fibre, shown in Figure 7.2. The optical fibre is a cylindrical structure with a central region, called the core,

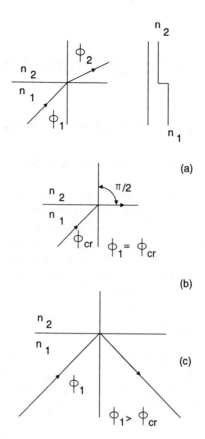

Figure 7.1 Principles of light guiding: (a) light refraction;
(b) the critical angle φ_{cr} for the limiting case of zero refraction;
(c) total internal reflection for $\varphi_1 > \varphi_{cr}$

and a surrounding external cylinder, called the cladding, with the
refractive index of the core being higher than that of the cladding. If
now a light ray is incident at the interface between core and cladding
at an angle greater than φ_{cr} then the light will propagate along the
fibre through a series of consecutive total reflections. This description
of the propagation of light assumes ideal interface conditions and any

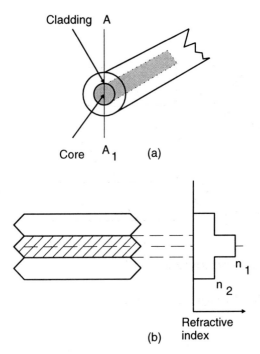

Figure 7.2   The geometry of a fibre waveguide: (a) physical shape; (b) the fibre internal geometry along the axis AA_1

departure from them will lead to both reflection and refraction and the light ray will lose energy as it is guided by the fibre.

Since the fibre guides light, one of the basic features is the ability to collect light from an optical source. This parameter is known as the numerical aperture of the fibre, NA, and in physical terms is half the apex angle of the light acceptance cone, as shown in Figure 7.3. It is given by Equation 7.3, where Δn is the difference in refractive index between core and cladding.

$$NA = n_o \sin \pi_a$$

$$= \left(n_1^2 - n_2^2 \right)^{1/2} \approx n_1 (2 \Delta n)^{1/2} \tag{7.3}$$

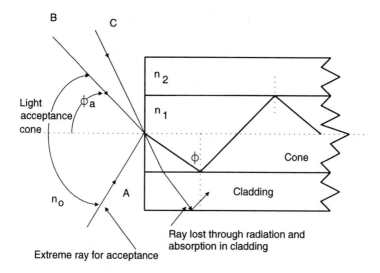

Figure 7.3   The light acceptance cone geometry

For instance if $\Delta n = 1\%$ and $n_1 = 1.5$ then NA = 0.212 and the light acceptance cone has an apex angle of 24.5°.

The propagation of light in a fibre can be understood at least qualitatively through the concepts of mode and that of the normalised frequency. An optical wave propagating by internal reflection can be represented as a bundle of rays, called modes, which is a concept of geometric optics. The concept of normalised frequency belongs to the formal analysis of the light propagation in fibre and is defined as in Equation 7.4.

$$V = \frac{2\pi}{\lambda} a \left(n_1^2 - n_2^2 \right)^{1/2}$$

$$= \frac{2\pi}{\lambda} a n_1 \left(2\Delta n \right)^{1/2} \tag{7.4}$$

Now the combination of the concepts of mode and normalised frequency enable us to derive a fundamental characteristic of light propagation in fibre. It was stated earlier that the fibre will accept and propagate all the rays entering the acceptance cone. The question now arises as to how many modes of propagation can be supported by the fibre. Assuming that the refractive index of the core is uniform then the number of modes is given by Equation 7.5.

$$M_{si} = 0.5 \left(\frac{2V}{\pi} \right)^2 \tag{7.5}$$

For example with $\Delta = 0.1$, $n_1 = 1.5$ and $a = 25\mu m$ the number of modes at $0.85\mu m$ is 311 and 936 at $1.55\mu m$. This regime of propagation is known as multimode propagation and implies that a number of rays propagate simultaneously. The alternative to multimode is single mode or monomode propagation, that is $M = 1$. The condition for single mode propagation is given by Equation 7.6.

$$V = \frac{2 \pi a \left(n_1^2 - n_2^2 \right)^{1/2}}{\lambda} < 2.045 \tag{7.6}$$

The parameter which must change, other things being equal, to make a multimode fibre single mode is the fibre radius. For the multimode fibre discussed above and for single mode propagation at $0.85\mu m$ the fibre radius should be reduced to $4.85\mu m$. However, this only shows the size of dimensions involved.

Usually the wavelength of operation is decided from other considerations and the single mode propagation conditions are satisfied by changing either Δn or a or both. Since very small Δn leads to high bend losses both parameters are used in designing for single mode propagation.

In order to preserve the mechanical strength of the fibre, the core diameter is reduced to around $8-10\mu m$ but the cladding is maintained at $60-125\mu m$. Since the standard multimode fibre has a core diameter of $50\mu m$ but a cladding diameter of $125\mu m$ it is not possible to identify the propagation conditions from a visual inspection of the fibre. One other restriction imposed by single mode propagation is that since only one mode propagates the optical source which excites the fibre must also be single mode.

It was assumed up to now that the refractive index of the fibre changes in a step-like way at the core-cladding boundary, as in Figure 7.4(a). This fibre is known as step index fibre and it is easy to make and is widely used. However, as it will be seen later, the information bandwidth of the step index multimode fibre can be improved if the refractive index of the core is graded with the maximum value at the centre of the core, as in Figure 7.4(b).

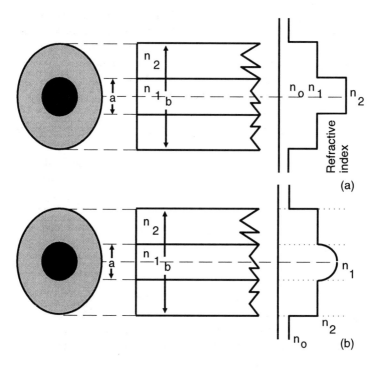

Figure 7.4   Fibre types: (a) step index profile; (b) graded index parabolic profile

Using axial symmetry in general the index variation can be written as in Equation 7.7, where α is the profile parameter.

$$n(r) = n_1 (1 - 2 \Delta n (r/a)^{\alpha})^{1/2} \quad r < a \quad (core)$$

$$n_2 = n_1 (1 - 2 \Delta n)^{1/2} \quad r \geq a \quad (cladding) \tag{7.7}$$

From the performance and manufacturing view point a very attractive profile is the one with $\alpha = 2$ known appropriately as the parabolic profile, shown in Figure 7.4(a). The change of refractive index affects the number of modes and for a parabolic profile the number of propagating modes is given by Equation 7.8.

$$M_{pp} = 0.5 \left(\sqrt{2} \frac{V}{\pi} \right)^2 \tag{7.8}$$

The number of modes therefore is halved. Single mode fibre are usually step index fibre because graded index single mode fibres do not offer any significant advantages.

The single and multimode fibres just outlined are made of silica or multicomponent glass compounds and dominate the communication market where performance is the prime consideration. For low bandwidth applications multimode plastic clad fibre or all plastic fibre can be used with significant cost reduction. The plastic clad fibre has plastic cladding but the core is silica. The all plastic fibre is of even lower cost but its performance is also lower.

7.1.2 Optical fibre characteristics

In order to use the fibre in communication systems the potential user needs to know their optical, electrical and mechanical characteristics. From the transmission point of view these can be reduced to essentially two: attenuation (loss) and dispersion (bandwidth).

The total loss in an optical fibre consists of two elements: intrinsic losses due to the fundamental properties of the materials used to make the fibre and extrinsic losses which are attributed to the actual fibre and the details of the usage.

The intrinsic losses are essentially two: absorption and scattering. Absorption is a property of the material and in optical wavelength it takes place when the energy of the incident optical radiation is absorbed by electronic transitions and dissipated through non-radiative processes. Fibre used in optical communications is made from glasses with high silica content doped with various oxides. For these glasses there are two absorption resonances. One is the ultraviolet and one in the infrared. At the ultraviolet part of the spectrum the loss is 1dB/km at 0.62μm but falls rapidly at longer wavelengths and at 1240nm it reaches 0.02dB/km. The infrared loss is around 0.03dB/km at 1500nm and 0.5dB/km at 1700nm.

If one assumes that 0.5dB/km loss is a reasonable one then there is a 1.5μm window between the two absorption edges. However, at this point scattering enters the picture. The scattering, known as Rayleigh scattering, is the result of microscopic material inhomogeneities. The inhomogeneities arise both from the glass and the various dopants which are used to produce the index difference, Δn, between core and cladding. Rayleigh scattering is sensitive to wavelength because it varies as λ^{-4}. Because of that scattering is insignificant at long wavelengths. Figure 7.5 shows the total loss of single mode fibre and it is interesting to notice that the global loss minimum of silica fibre is at 1550nm where the loss attains the minimum value of 0.18dB/km.

The extrinsic fibre losses owe their presence to the practical difficulties in making and using fibres. Extrinsic losses include scattering due to the departure from perfect core-cladding interface, bending losses, losses due to the permanent joining of fibre (splicing), loss of connectors and coupling losses at the input and output. The scattering at the core-cladding interface results in both mode conversion and radiation loss. This loss can be controlled by increasing the radius of the core and the index difference, Δn. Bending loss depends on the size of the core and the bend radius. The bending loss can be represented as in Equation 7.9.

$$\alpha_{bl} = C e^{-R/R_c} \tag{7.9}$$

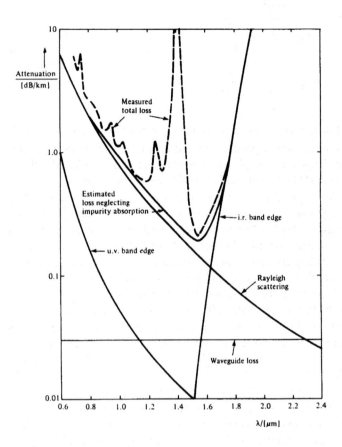

Figure 7.5 Loss of single mode fibre

R is the radius of curvature of the bend and R_c the critical curvature given by Equation 7.10, with a being the fibre radius.

$$R_c \approx \frac{3\, n_1^2\, \lambda}{4\, \pi \left(\; n_1^2 - n_2^2 \;\right)^{3/2}}$$

$$\approx \frac{a}{2\, n_1\, \Delta n} = \frac{a}{N A^2} \tag{7.10}$$

Looking at the equation for R_c the approach to minimising α_{bl} is to use the largest possible Δn and operate at the shortest possible wavelength. This implies that operation at long wavelengths, that is around 1550nm, will possibly lead to higher bend losses.

The physical cause of bend losses is the loss by radiation of the energy, since a tight bend forces the radiation to propagate into the cladding. The bending losses introduced and discussed up to now were due to gross fibre bends introduced by the user. However, there are bending losses introduced randomly along the waveguide, with each of them introducing a very small loss but the aggregate loss can be quite high. This is known as microbending loss and it can be caused by pressure exerted by other fibre within a fibre cable.

One additional form of loss is the result of the small variation in the core diameter of a fibre arising during the manufacturing phase. The physical mechanism is similar and the loss is known as waveguide loss.

Fibre can be pulled in lengths between a few metres to a few kilometres and since system length range from a few kilometres to tens of kilometres, permanent joint techniques are required. The permanent jointing fibre is known as splicing. Two broad categories of splicing techniques have emerged: fusion splicing (welding) and mechanised splicing.

Fusion splicing is accomplished by the fusion of two pre-aligned fibres by the application of heat. The heat can be generated by an electric arc or flame, and the technique is easy to apply and yields remarkable low losses in the field. For example fusion splicing yields routinely 0.2dB loss per splice. One drawback of fusion splicing is the reduction of the tensile strength of the fibre in the vicinity of the fused splice because of the heating of the fibre. For this reason fusion splices are packaged with the aim of reducing the tensile load in the vicinity of the splice.

Mechanical splicing, before the introduction of fusion splicing, was the only way of permanently joining two fibres. The basic idea behind this mechanical splicing is the alignment of two fibre ends and then through the use of a transparent adhesive with refractive index which matched that of the fibre, the joint is made permanent. The alignment of the fibre and the mechanical strength are achieved by using capillaries or a V-groove. The performance obtained with mechanical splicing is in the range of 0.2dB to 0.3dB. Which splicing technology is used depends on the application but fusion is simpler to use and it is very popular.

In optics the term dispersion is used to indicate the dependence of the refractive index of the material on the wavelenght of the radiation. This dependence is a basic characteristic of the material. The effect of dispersion on a pulse of light propagating in the fibre is to increase the width of the pulse and the understanding of dispersion and its effects is fundamental in understanding optical fibre transmission.

Optical energy propagates along the waveguide with a velocity known as the group velocity, u_g, given by Equation 7.11 where, in this equation, the denominator is known as the group index.

$$u_g = \frac{c}{n_1 - \lambda \left(\dfrac{d\,n_1}{d\,\lambda} \right)} \tag{7.11}$$

Clearly if $\dfrac{d\,n_1}{d\,\lambda} = 0$ then the group velocity is independent of wavelength and is equal to the phase velocity. An optical pulse consists of a number of wavelengths (frequencies) and since each of them propagates with different group velocity they arrive at the end of a fibre of length L each delayed by τ_g given by Equation 7.12.

$$\tau_g = \frac{L}{u_g} \tag{7.12}$$

For normal dispersion the higher frequencies are delayed more than the lower ones, with the result that the pulse width increases. Assuming linear operation and that the output pulse emerging from the waveguide is Gaussian of standard deviation (r.m.s. value) σ then the 3dB optical bandwidth is given by Equation 7.13.

$$B\,W_{opt} = \frac{0.1874}{\sigma} \tag{7.13}$$

The 3dB electrical bandwidth is defined as the frequency at which the optical power is down to 0.707 of that at zero frequency, as in Equation 7.14.

$$B\,W_{elc} = \frac{0.1325}{\sigma} \tag{7.14}$$

After this brief introduction to the concept and impact of dispersion we proceed to examine the effect of dispersion in single and multimode fibres.

7.1.2.1 *Dispersion*

The dispersion of a fibre of arbitrary parameters can be separated into two elements, intramodal and intermodal. The intramodal dispersion is the dispersion whose origin lies in the interaction of the material of the waveguide and of the waveguide itself with the optical signal. It consists of two main components, material or chromatic dispersion and waveguide dispersion. The origin of the intramodal dispersion, known also as chromatic dispersion, is the finite spectral width of the optical source and its interaction with the waveguide material properties (material dispersion) and the structure (waveguide dispersion).

Intermodal dispersion, also referred to as modal dispersion, is the result of the propagation delay between modes in a multimode fibre. Both types of dispersion are present in a multimode fibre because each mode is subject to intramodal dispersion but as should be clear by now only intramodal dispersion is present in a single mode fibre. Since intramodal dispersion is more fundamental we examine it first.

Consider a pulse of near monochromatic radiation of average wavelength propagating in a single mode fibre of length L. Then combining Equations 7.11 and 7.12 the pulse delay due to the material dispersion is given by Equation 7.15.

$$\tau_d = \frac{L}{u_g} = \left(\frac{L}{c} \right) \left(n_1 - \lambda \left(\frac{d\,n_1}{d\,\lambda} \right) \right) \tag{7.15}$$

Assuming that the r.m.s. spectral width of the radiation is σ_s then the pulse broadening due to the material dispersion is given by Equation 7.16.

$$\sigma_m = -\left(\frac{L\,\sigma_s}{c} \right) \left(\lambda \frac{d^2 n_1}{d\,\lambda^2} \right) \tag{7.16}$$

The material dispersion parameter, $D\,(\lambda)$ is usually defined as in Equation 7.17, which gives Equation 7.18.

$$D\,(\lambda) = -\left(\frac{\lambda}{c} \right) \left(\frac{d^2 n_1}{d\,\lambda^2} \right) \quad ps/nm.km \tag{7.17}$$

$$\sigma_m = \sigma_s D\,(\lambda)\,L \tag{7.18}$$

The physical origin of waveguide dispersion is the non-linear dependence of the propagation constant of the waveguide on the optical frequency. This is the irreducible minimum dispersion and it is present in the waveguide even for zero material dispersion. The waveguide dispersion can be approximated by Equation 7.19.

$$\Delta\,\tau_w = \frac{L}{c}\,n_1\,\Delta\,n\,V\,\frac{d^2\,(V\,b)}{d\,V^2}\,\frac{\Delta\,\lambda}{\lambda} \tag{7.19}$$

For V in the range $1.5 < V < 2.4$ the parameter b can be approximated by Equation 7.20.

$$b = \left(1.428 - \frac{0.996}{V} \right)^2 \qquad (7.20)$$

Since for single mode transmission $2.0 < V < 2.4$ and $d^2(Vb)/dV^2$ is of the order 0.1 to 0.2. At short wavelengths, that is around 1μm, both waveguide and chromatic dispersion have the same sign and they are added.

As it was stated above the material dispersion for silica decreases to zero around 1.28μm. At wavelengths longer than that the waveguide and material dispersion are of opposite sign and as a result the wavelenght of zero dispersion moves to longer wavelengths. In general waveguide dispersion is small and by itself a second order effect, but by a suitable choice of fibre parameters it can be made to cancel the material dispersion yielding a fibre with zero dispersion at longer wavelength than the natural zero dispersion of the silica. It is particularly important that the dispersion zero can be moved to 1550nm to coincide with the global loss minimum of the silica fibre. The total dispersion of a single mode fibre is shown in Figure 7.6.

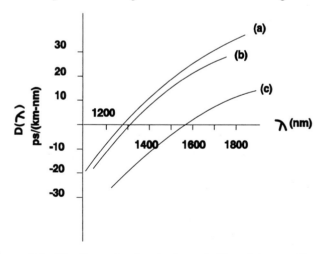

Figure 7.6   The dispersion for single mode fibre: (a) pure silica; (b) non-dispersion shifted fibre (NDSF); (c) dispersion shifted fibre (DSF)

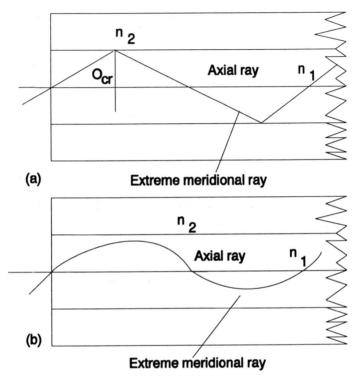

Figure 7.7 Propagation of extreme rays: (a) in step index fibre; (b) in parabolic refractive index fibre

The dispersion of a multimode fibre consists of the material and waveguide dispersion of a single mode fibre and of the intermodal dispersion. The nature of the intermodal dispersion can be understood by considering a step index multimode fibre.

The light launched into this fibre can propagate in the extremes along two paths as in Figure 7.7(a). The first is along the axis of the fibre and along this path the pulse will be subjected to the minimum delay, ΔT_{min}, given by Equation 7.21.

$$\Delta T_{min} = \frac{L}{c/n_1} \tag{7.21}$$

The second is the extreme meridional ray which will experience the maximum delay, given by Equation 7.22.

$$\Delta T_{max} = \frac{L/\cos\theta}{c/n_1} = \frac{L\,n_1}{c\cos\theta} \tag{7.22}$$

Using Snell's law of refraction gives Equation 7.23.

$$\Delta T_{max} = \frac{L\,n_1^2}{c\,n_2} \tag{7.23}$$

The delay difference $\Delta\tau_d$ between the two rays is given by Equation 7.24.

$$\Delta\tau_d = \Delta\tau_{max} - \Delta\tau_{min} = \frac{L\,n_1^2}{c\,n_2}\left(\frac{n_1 - n_2}{n_1}\right)$$
$$= \frac{L\,n_1}{c}\,\Delta n \tag{7.24}$$

Using the numerical aperture of the fibre gives Equation 7.25.

$$\Delta\tau_d = \frac{L\,(NA)^2}{2\,c\,n_1} \tag{7.25}$$

Using $n_1 = 1.5$ and $\Delta n = 0.01$ we obtain $\Delta\tau_d = 50$ps/km. Comparing this to the material dispersion of a single mode fibre around 1.3μm the difference is substantial and the question is as to whether or not the dispersion of a multimode fibre can be reduced. Before we discuss this issue it will be very important to see what happens in practice.

The maximum dispersion (see Equation 52.25) is usually reduced in practice through a number of mechanisms. The first and obvious one is that high order modes, that is those with the maximum delay, exhibit higher losses because they penetrate deep in the cladding. The second and not so obvious is mode mixing. Because of the irregularities at the core-cladding interface of an actual fibre energy is transferred from slower to faster modes and vice versa, with the result that in a long fibre the intermodal dispersion is reduced. With mode coupling between all guided modes the dispersion is given by Equation 7.26, where L_m is the characteristic length for mode mixing.

$$\Delta \tau_s = \frac{n_1 \Delta n}{c} (L L_m)^{1/2} \tag{7.26}$$

In practice L_m is established by measurements.

In spite of the effect of mode mixing in reducing intermodal dispersion the most successful approach in reducing the dispersion of multimode fibres is by grading the refractive index of the core. The most popular grading profile is the parabolic which yield a dispersion given by Equation 7.27.

$$\Delta \tau_s = \frac{L n_1 \Delta n^2}{8 c} \tag{7.27}$$

The pulse broadening given by Equation 7.26 or 7.27 corresponds to the maximum value and if used in system design it gives the worst case. Since in practice the value obtained is smaller but depends on a number of random parameters it is acceptable to use the root mean square value of the pulse broadening. Assuming that the light is launched into the fibre uniformly over the cone of acceptance the r.m.s. pulse broadening for step index and parabolic graded index fibre is given by Equation 7.28.

$$\sigma_{si} \approx \frac{L (NA)^2}{4 \sqrt{3} \, n_1 c} = \frac{L n_1 \Delta}{2 \sqrt{3} \, c}$$

$$\sigma_{gi} = \frac{L n_1 \Delta n^2}{20 \sqrt{3} \, c} \tag{7.28}$$

The total dispersion of the multimode fibre is given by the sum of the intramodal and intermodal dispersion but in the mean square sense. The reason for this lies in the independence of the physical effects that cause the pulse broadening. The total dispersion, σ_t, is given by Equation 7.29, where σ_m and σ_d are the chromatic (intramodal) and intermodal pulse broadening respectively.

$$\sigma_t = (\sigma_m^2 + \sigma_d^2)^{1/2} \tag{7.29}$$

The total fibre dispersion can be used to establish an important figure of merit for a fibre, the bandwidth-length product, $B_{opt}L$. This is given by Equation 7.30, where σ_t refers to the dispersion per unit length, usually the kilometre.

$$B_{opt}L = \frac{0.187}{\sigma_t} \approx \frac{0.2}{\sigma_t} \tag{7.30}$$

The performance of a state of the art single mode fibre regarding losses and dispersion is summarised in Figures 7.5 and 7.6. The loss in a state of the art multimode fibre is similar to that of a single mode with 2.0dB/km to 2.5 dB/km at 850nm and 0.55dB/km to 0.6 dB/km at 1300nm. The difference of course lies in bandwidth. The bandwidth of graded index 50/125µm multimode fibre is around 600MHz.km and 1500MHz.km at 850nm and 1300nm respectively. The equivalent r.m.s. dispersion is 333.3ps/km and 125ps/km respectively and it is easy to see the advantage of single mode transmission.

7.2 Optical sources

7.2.1 The basics of light generation

Matter absorbs or emits radiation by electron transitions between energy levels which characterise the material. According to quantum mechanics this interaction takes place only in discrete quantities of energy. That is assuming that two energy levels E_1 and E_2 exist with $E_2 > E_1$ then the minimum energy which can be absorbed or emitted is E_g (where $E_g = E_2 - E_1$) which corresponds to a characteristic wavelength λ_c given by Equation 7.31, where h is Planck's constant

$(6.6261 \times 10^{-34} \text{Js})$, c is the speed of light $(2.999 \times 10^8 \text{m/s})$ and λ the wavelength of the radiation.

$$\lambda_c = \frac{h\,c}{E_g} = \frac{1.24}{E_g\,(e\,V)} \quad (\mu m) \tag{7.31}$$

There are three ways in which absorption or emission can take place. Consider again the two level system, shown in Figure 7.8, but

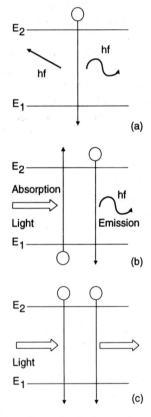

Figure 7.8   The interaction of radiation and matter: (a) random emission; (b) absorption and random emission; (c) stimulated emission and amplification

with electrons at level E_2. Then without external light irradiating the material the electrons will drop spontaneously to level E_1 emitting photons of energy E_g. This emission is completely random and without any triggering mechanics being involved.

Consider now that the material is irradiated with light of wavelength λ_o close to λ_c. Then the incoming photons will be absorbed by the material and electrons will now appear at level E_2. The electrons stay at E_2 for a time interval $\Delta \tau$ whose duration is a characteristic of the material. If now the external irradiance is removed the electrons will delay to level E_1 through spontaneous emission and eventually the level E_2 will be empty.

The third possibility is the most important. Consider again the two level system but this time there are electrons at level E_2 and the material is irradiated. Then the electrons at level E_2 interact with the radiation, drop to level E_1 and photons are released in phase with those of the input light. This phenomena is called stimulated emission and under certain conditions can lead to light amplification. This is a fundamental concept and its understanding is invaluable in understanding the operation of optical sources.

Consider again the two level system with a total number of atoms N of which n_1 are at E_1 and n_2 at E_2. Under thermal equilibrium the number of atoms at level E_2 is given by Equation 7.32, where k is Boltzmann's constant $(1.38 \times 10^{-23} \mathrm{JK^{-1}})$ and T the absolute temperature in Kelvin.

$$n_2 = n_1 \exp\left(\frac{-hf_o}{kT} \right) \tag{7.32}$$

Since $T > 0$ then $n_1 > n_2$. If now external light irradiates the material the probabilities of absorption $(E_1 \rightarrow E_2)$ or emission $(E_2 \rightarrow E_1)$ are proportional to the intensity of light and the number of atoms at levels E_1 and E_2, that is, the probability of absorption is approximately equal to In_1 and the probability of emission is approximately equal to In_2.

Since $n_1 > n_2$ the material absorbs the radiation. Now consider that somehow $n_2 > n_1$ but since the probability of absorption is as before emission predominates and we have achieved Light Amplification by

Stimulated Emission of Radiation (Laser). The condition that leads to laser action is known as population inversion.

Using the basic mechanisms of matter radiation interaction two classes of optical sources have emerged. Devices using spontaneous emission (Light Emitting Diodes or LEDs) and devices using light amplification (lasers). The details of their operation depend on the material systems used and we briefly discuss this important issue.

Spontaneous emission and laser action has been observed in a large number of material systems ranging from glasses to crystals to semi-conductors. Which material system is used depends on a number of requirements imposed by the application and for fibre communications the indisputable choice is that of semiconductor materials. Optical sources made of semiconductors offers excellent perfor-mance both optical and electronic, small size, reliability and perhaps what may be the most important feature for future evolution the possibility of integrating optical sources with the electronic functions using the same material system.

Since the emitted frequency is a function of the energy bandgap a number of semiconductor material systems have evolved to satisfy the transmission requirements. The first system to be developed was the ternary $Ga_{1-x}Al_xAs/GaAs$ ($0 \leq x \leq 1$). The subscript x is referred to as the mole fraction and by varying it one varies the wavelength at which the device operates.

The extremes of the operation are obtained with $x = 0$, $\lambda = 0.9\mu m$ and with $x = 0.4$, $\lambda = 0.65mm$. The GaAlAs system therefore covers the $0.8\mu m$ fibre window, the first optical window to be used. A study of the fibre loss verses wavelengths will highlight the simple fact that operation of $0.8\mu m$ is not the optimum.

The silica fibre has the zero dispersion of around $1.3\mu m$ and the global loss minimum at $1.55\mu m$. To operate at these two windows a quarternary compound semiconductor has been developed, $In_xGa_{1-x}As_yP_{1-y}$.

Now there are two mole parameters and the bandgap energy corresponds to wavelength ranging from $0.92\mu m$ to $1.65\mu m$. This material therefore covers both the $1.3\mu m$ and $1.55\mu m$ windows. Other materials have been investigated but they have never reached the maturity of the two outlined above.

7.2.2 Light emitting diodes and lasers

The operation of LED and lasers is based on the same principle, that of population inversion (negative temperature), but the structures used are radically different.

Population inversion is obtained in a semiconductor p-n junction by heavily doping both types of the material. This leads to a degenerative p-n junction which means that at equilibrium this Fermi level lies within the valence and conduction bands and the state between the Fermi level and the end of the band are filled.

With strong forward bias there exist at the depletion layer a region which is doubly degenerate, that is contain degenerate electron and holes. This corresponds directly to the population inversion condition. In this region, called the active region, electrons are available for either spontaneous or stimulated emission. Which of the two predominates depends on the structure.

An LED relies on the spontaneous emission for operation and the structure of typical devices are shown in Figures 7.9 and 7.10. The

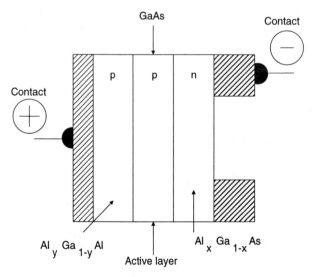

Figure 7.9 Double heterojunction LED

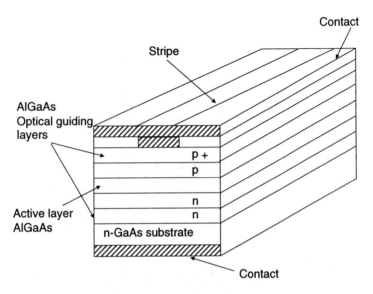

Figure 7.10   Structure of an edge emitting LED

device consists of three layer two $Al_xGa_{1-x}As$ and one GaAs (x = 0). Since the bandgap of AlGaAs is larger than that of GaAs the spontaneous emission generated in the active layer (GaAs) is not absorbed and it is available for collection and launching into a fibre. This structure is known as double heterojunction because of the p-p and p-n junction.

Devices operating on the principle of stimulated emission require, in addition to population inversion, the presence of radiation. In a laser this 'radiation' is obtained through optical feedback. Optical feedback is established by embedding the active material into a cavity whose walls are mirrors, (Figure 7.11). Then the operation is like that of any other oscillator. When heavy forward bias is applied the population is inverted and spontaneous emission follows. Some of the photons of the spontaneous emission are reflected from the end mirrors and playing the role of the radiation are irradiating the active material. Due to the stimulated emission effect the reflected radiation is amplified and laser action commences.

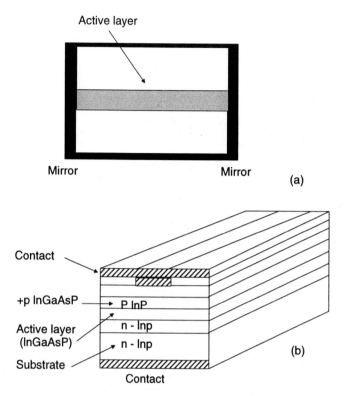

Figure 7.11   Principles of optical feedback: (a) schematic of a laser; (b) structure of a long wavelength laser

From both the theoretical and practical points it is extremely important to know the optical wavelength (frequency) at which LEDs and lasers radiate. With LED the situation is simple. The emitted radiation corresponds to the spontaneous emission. This can be quite broad 30µm to 50µm and it is centred at the λ_c. For the laser the situation is very different. Being an oscillator there are two conditions for operation: internal gain to overcome losses in the resonator and positive feedback.

The gain, which corresponds to the range over which the material will amplify is characteristic of the material. The cavity however

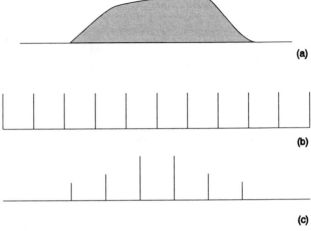

Figure 7.12   The generation of the discrete laser spectrum: (a) device gain; (b) cavity modes; (c) laser spectrum

imposes additional conditions through the feedback mechanism and selects the frequencies at which the device oscillates. Figure 7.12 shows the basics of the operation. The device gain is $g(\lambda)$. The frequencies at which the device can potentially oscillate are those for which there is positive feedback. These wavelengths are called the longitudinal modes of the cavity. The spacing of the modes is given by Equation 7.33, where n is the refractive index of the material and is L the length of the device.

$$\Delta\lambda = \frac{\lambda^2}{2nL} \tag{7.33}$$

The number of the longitudinal modes which can be supported by the cavity is given by Equation 7.34.

$$M_{lg} = 2n\frac{L}{\lambda} \tag{7.34}$$

For example at $\lambda = 0.850\mu m$ for GaAs with n = 3.5 and device length, L, of say $200\mu m$ the number of modes is 1647. However, only

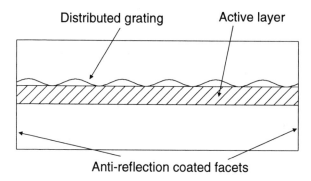

Distributed grating   Active layer

Anti-reflection coated facets

Figure 7.13   Schematic of a DFB laser

those modes for which the device has gain can oscillate and, in contrast to an LED spectrum, which is continuous the laser spectrum is discrete. For the device used in the example, the mode spacing is $\Delta\lambda = 0.516$nm. Depending on the design and driving conditions the number of laser modes range from one (single mode) to many (multi-mode). 'Many' usually implies 5 to 10 modes.

In semiconductor lasers the mirrors are made by simply cleaving the device to the required length and leaving the facet un-coated. The laser devices that depend for feedback on two mirrors on the facets of the active layer is known as a Fabry Perot laser. This is the simplest possible laser structure and naturally it has a number of disadvantages especially mode instability. To cure this problem advanced laser structures, such as distributed feedback lasers (DFB), are being introduced. The optical feedback for these devices is provided not by mirrors at the end of the cavity but by a grating built into the laser which provides continues feedback along the whole length of the device, as in Figure 7.13. The result is virtual single mode lasers because the other modes are 25–30dB below the fundamental.

7.2.3 Device characteristics

The characteristics required to describe the operation of optical sources depend on the application. For fibre communication applications a minimum set of parameters which can describe the device operation satisfactory is as follows:

LEDs: light output vs current, spectral density, information bandwidth.

Lasers: light current characteristics, spectral density, turn on delay, relaxation oscillation, spectral broadening under modulation, partition noise and information bandwidth.

The light current characteristics of a LED is shown in Figure 7.14 for three device structures. Notice that for the surface and edge emitting LEDs the device exhibits a linear characteristic for low currents with the edge emitting device being linear for a substantial part of the characteristic. The spectral density of a LED is continuous, Gaussian shaped and quite broad. Depending on the details of the structure it ranges from 25nm to 40nm for the 850nm window and

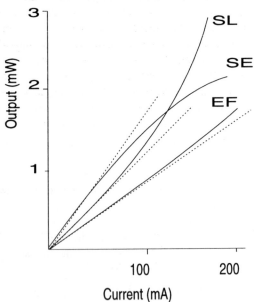

Figure 7.14   The light-current characteristic of LEDs

from 50nm to 120nm for the 1.2μm to 1.7μm range. The spectral width quoted applies for the surface emitting (SE) structure. The spectral width of an edge emitting (EE) device is 0.6 that of a SE and that of a superluminence (SL) about 0.25 of the SE spectral width.

The information bandwidth of LEDs is given by Equation 7.35, where the constant τ is the total carrier lifetime.

$$P(\omega) = \frac{P(O)}{(1 + (\omega\tau)^2)^{(1/2)}} \tag{7.35}$$

The modulation bandwidth, f, is given by Equation 7.36.

$$\Delta f_{LED} = \frac{\Delta\omega}{2\pi} = \frac{1}{2\pi\tau} \tag{7.36}$$

Since the lifetime depends on the current density the bandwidth is a function of the bias of the device and it drops as the power increases.

There are a fair number of LED device but the typical parameters of devices suitable for fibre communications are summarised in Table 7.1. As one can see from this table the output power into a standard multimode fibre tail is of the order of a few tens of microwatts only in spite of the fact that the power from the facet of the device is of the order of 5mW to 15mW. The reason for this difference is the broad far field radiation pattern of the LEDs and the relatively small NA of the standard multimode telecommunication fibre.

Table 7.1 Typical LED performance parameters with multimode fibre tail (50/120μm)

λ_p (nm)	$\Delta\lambda$ (nm)	P_0 (μW)	BW (MHz)	t_{rf} (ns)	C_t (pF)	V_F (V)	I_b (mA)
865	45	80	200	10	220	2.0	100
1300	80	20	400	2	300	1.5	100
1550	125	20	—	3	—	1.8	150

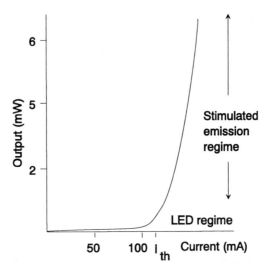

Figure 7.15   The light-current characteristic of a laser

The light current characteristic of a laser is shown in Figure 7.15. In contrast to the same curve for LEDs there is now a sharp 'knee'. The current corresponding to the knee is the threshold current, I_{th}, and it is an important parameter of the device. At currents below I_{th} the laser behaves like an LED, that is, the radiation is spontaneous and the power low. Above I_{th} the power rises very rapidly and now the device operates in the regime of stimulated emission. The spectral density of the laser is now very different from that of an LED. It is narrow and consists of a number of spectral lines under an envelope that is nearly Gaussian. The spectral density depends on the driving conditions of the device especially with large digital or analogue signals.

One special feature of the laser is the turn on delay. With the device biased at an arbitrary current level, I_b, below threshold, the time taken by the optical radiation to rise as a result of an applied pulse is the turn on delay t_d as shown in Figure 7.16, and is given by Equation 7.37, where τ is the recombination time constant and I the applied signal current.

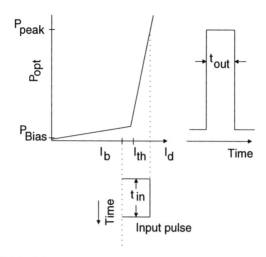

Figure 7.16   The concept of turn on delay in semiconductor lasers: $t_{out} < t_{in}$ for $I_b < I_{th}$

$$t_d = \tau \ln \left(\frac{I - I_b}{I - I_{th}} \right)$$
(7.37)

For $I_b \rightarrow I_{th}$, $t_d \rightarrow 0$ with the maximum of t_d obtained for $I_b = 0$. Biasing the device at or above threshold, the turn on delay is zero. This implies that for high bit rate systems pre-bias is mandatory, otherwise there will be a pattern effect. In addition to the turn on delay a laser driven by a digital signal exhibits relaxation oscillations. Clearly the period of the relaxation oscillations should be less than the width of the pulse otherwise there will be again a pattern effect.

It was mentioned above that the spectral density of a laser depends on the modulation. This leads to two related phenomena, spectrum broadening and partition noise. Spectral broadening manifests itself with the appearance of new spectral lines which where not there under CW driving conditions. The cause of partition noise is the variability of the energy of each laser mode as the device is modulated. Now, assuming that the pulse propagates over a dispersive medium, each mode suffers different delay and at the receiver the

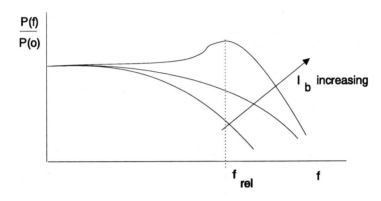

Figure 7.17 Small signal frequency response of semiconductor lasers

pulse amplitude fluctuates, suggesting the existence of noise process in operation. The information bandwidth of lasers range for a few megahertz to gigahertz with current experimental devices reaching 20GHz. The theoretical bandwidth relation is as in Equation 7.38.

$$P(\omega) = P(O) \frac{\omega_o^2}{(\omega_o^2 - \omega^2) + j\beta\omega} \qquad (7.38)$$

In this equation ω and $\beta\omega$ are given by Equations 7.39 and 7.40, and τ_{sp} and τ_{ph} are the spontaneous recombination time and mean photon lifetime respectively, as in Figure 7.17. Again as in LEDs the bandwidth depends on bias current.

$$\omega_o^2 = \frac{(I_o - I_{th})}{\tau_{sp}\,\tau_{pt}\,I_{th}} \qquad (7.39)$$

$$\beta = \frac{I_o}{\tau_{sp}\,I_{th}} \qquad (7.40)$$

Table 7.2   Typical laser performance parameters with single and multimode fibre tails for digital applications

(1) corresponds to DFB laser

λ_{peak} (nm)	I_{th} (mA)	P_{out} (mW)	$\Delta\lambda$ (nm)	t_r (ns)	t_f (ns)	Device class	Fibre tail
(nm)	(mA)	(mW)	(nm)	(ns)	(ns)	Class	Tail
1550	20	1.3	see(1)	0.1	0.1	DFB	S/M
1300	20	2.0	see(1)	0.1	0.1	DFB	S/M
1550	30	2.0	4.0	0.5	0.7	F-P	S/M
1300	40	1.5	5.0	1.0	1.0	F-P	S/M
850	16	4.0	3.0	0.4	0.4	F-P	M/M
1300	20	3.0	8.0	0.3	0.3	F-P	M/M

At $\omega = \omega_o$ the device exhibits a resonance which corresponds to the relaxation oscillation frequency.

It was mentioned above that multimode lasers are subject to spectral problems. The solution to these problems lie in operating the device in a single longitudinal mode. Fabry Perot devices cannot generally operate in a stable single mode. As was mentioned earlier a successful single mode device is the Distributed Feedback laser (DFB). These devices are used in all long haul systems operating at 1550nm with or without dispersion shifted fibre. However, even these devices are not free of imperfections. They suffer from chirping, that is the small change of wavelength during the duration of the current pulse corresponding to a digital '1' symbol.

The power from the facet of lasers designed for fibre communication is around 5mW to 10mW but the power available from the fibre tail is of the order at 1mW to 2mW. This again is due to the rather broad and asymmetric far field of the lasers.

The performance of lasers varies depending on the application but for fibre communications Table 7.2 summarises the characteristics of some typical lasers.

Table 7.3   Typical laser performance parameters with single and multimode fibre tails for analogue applications

(1) is in dBc units

λ_{peak} (nm)	I_{th} (mA)	P_{out} (mW)	$\Delta\lambda$ (nm)	BW (GHz)	2nd H. Dist. (dbm)	3rd Or. Int. (dbm)	Fibre tail
1300	20	2.0	8	15.0	−15	−30	M/M
1300 1550	25	2.0	7.0	1.5	−40[1]	−50[1]	M/M
1310 1550	20	4	—	0.6	−60[1]	−65[1]	S/M

The performance of some lasers for analogue modulation is summarised in Table 7.3.

These tables contain but a fraction of what is available on semiconductor lasers and the interested reader is advised to consult manufacturer's catalogues for further information.

7.3 Optical detectors

7.3.1 The basics of optical detection

The function of an optical detector is to recover the information which was imposed on the optical carrier. There are many effects which can be used for optical detection but in terms of performance detectors based on the photoelectric effects and using semiconductor material outperform all others, at least in the field of fibre communication, and we will concentrate on them.

The principle of photodetection is simple. Photons are absorbed by the material and electron from the valence band are raised in the conduction band leaving behind holes. If an electric field is applied across the device both electrons and holes are swept out of the device creating the photocurrent.

Since the generated photocurrent is proportional to the intensity of the irradiance, that is the square of the optical field, the detection process is also known as direct detection and is the most widely used form of optical detection. Mathematically it is stated as in Equation 7.41, where R is a constant to be discussed later.

$$i_{ph}(t) = R e(t)^2 = R P(t) \tag{7.41}$$

The most successful detector structure has been the p-i-n photodetector, shown in Figure 7.18. The device consists of three layers. Two heavily doped layers, p^+ and n^+, and a lightly doped n^- layer sandwiched between the two. The n^- layer is so lightly doped that it is virtually an intrinsic semiconductor. Hence the name p-i-n.

In order to minimise the capacitance, the device operates reversed biased. The operation is very simple. The radiation penetrates the thin p^+layer with minimum absorption and it is totally, or nearly totally, absorbed in the i layer. The threshold wavelength for absorption is given by the Equation 7.42, which is the same equation met in the interaction of radiation with matter, in the section on optical sources.

$$\lambda_{th} = h \frac{c}{E_g} = \frac{1.24}{E_g(eV)} \quad (\mu m) \tag{7.42}$$

Photodetectors are transducers since they convert optical power into electrical power. A critical parameter of a detector is the efficiency with which it performs this conversion. This parameter is known as the quantum efficiency and is defined as in Equation 7.43.

$$m_q = \frac{electrons\ collected}{electrons\ available} = \left(\frac{I_{ph}/e}{P_{opt}/hf} \right) \tag{7.43}$$

The value of n_q with wavelength depends on the detector material and is a function of wavelength. Figure 7.19 shows the absorption of some typical detector materials and for fibre communications Si, Ge

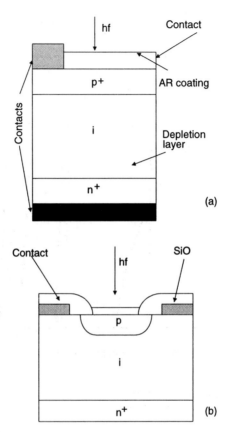

Figure 7.18 The p-i-n photodetector: (a) the schematic of a p-i-n; (b) silicon p-i-n photodetector

and the quarternary/ternary InGaAsP cover the wavelength range of interest. Using Equation 7.41 the photocurrent can now be written as in Equation 7.44, where e is the electronic charge and R the device responsivity in [A/W].

$$i_{ph}(t) = n_q e \frac{P(t)}{hf} = R P(t) \tag{7.44}$$

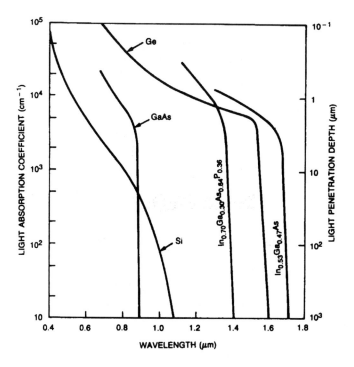

Figure 7.19 The optical absorption coefficient for some materials used in fibre optic communications

The principle of the operation of the p-i-n photodetector is that in the limit as $n_q \rightarrow 1$ one photon produces one electron-hole pair. However, for reasons which will become apparent in the section on receivers, if the device could yield for each photon many electron-holes pairs then the impact of the noise of the electronic amplifiers following the detector on the total (S/N) ratio will be minimised. Such devices, known as Avalanche Photodetectors (APDs), make possible the design of high sensitivity optical receivers.

In an APD, electron-hole pairs released by the primary photoelectron effect are accelerated to velocities capable of initiating impact ionisation. The secondary pairs, who owe their presence to ionisation, in their turn generate other pairs and the final current is many times larger than primary photocurrent. The APD designs are more complex than those for p-i-n detectors but essentially a p-i-n detector can function as an APD with the i-layer both absorbing the photons and accelerating the photoelectrons and holes. Of course such a device is not as good as could be made and special designs have evolved for APDs. The current in the output of an APD is given by Equation 7.45 where M is the current gain of the device.

$$I_{apd}(t) = i(t)_{ph} M = n_q \frac{P_{opt}(t)}{h\upsilon} M \qquad (7.45)$$

Clearly for M = 1 the output current equals the photocurrent as in a p-i-n detector. The maximum usable value of M depends on the material, the device structure and the quality of the semiconductor processes used to produce the device.

Typical Si and InGaAsP designs as shown in Figure 7.20. The Si device, known as reach through APD, offers good quantum efficiency down to 850nm in spite the fact that the Si absorption starts tailing off. The InGaAsP device is specially designed for long wavelengths in the sense that the absorption and multiplication do not take place in the same layer. Such devices, known as separate absorption and multiplication detectors (SAMs for short), aim at minimising the multiplication noise.

7.3.2 The characteristics of optical detectors

The device parameters of a p-i-n detector relevant to the design of an optical receiver are the quantum efficiency n_q, or responsivity R, the capacitance C_d, the series resistance R_s, the dark current I_d and the device bandwidth. The n_q or R depend on the material and the device design. For well designed devices n_q of the order of 80% to 95% is possible. Since p-i-n detectors operate reversed biased the device capacitance can be approximated to that of a plane capacitor, given by Equation 7.46 where ε_o is the free space permittivity, εr the relative permittivity, A_d the device junction area and w the width of the i-layer.

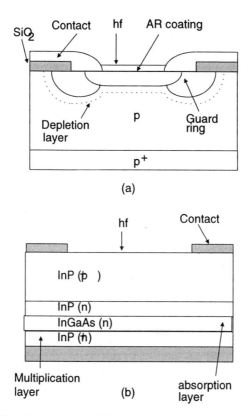

Figure 7.20   Some typical APD device structures: (a) silicon; (b) InGaAsP

$$C_d = \varepsilon_o \, \varepsilon_r \, \frac{A_d}{w} \tag{7.46}$$

The value of C_d depends on the material (ε_r), the device area (A_d) and the width of the i-layer width. For fast devices with 50μm diameter the capacitance can be as low as 0.5pF. The series resistance is the resistance of the material and of the contact bonding. In well designed detectors the series resistance is of the order of 5 to 20 ohms.

The dark current of the device is the current flowing in the circuit with bias applied but without light. This current is small because the device is reversed biased but it depends on temperature, device structure and material quality. In general devices for long wavelengths (because of the smaller band gap) have larger dark currents than those for short wavelength devices. For example Si p-i-n have much lower I_d that both Ge and InGaAsP, pAs rather that nAs. The device operates reversed biased and consequently the device resistance, that is the resistance of the junction, is large and is usually ignored in designs. The equivalent small signal circuit for normally biased photodiodes is shown in Figure 7.21.

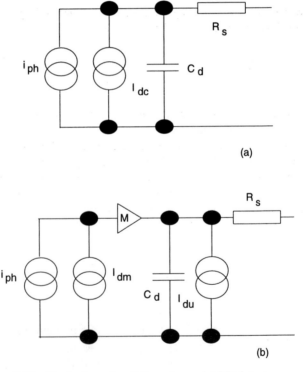

(a)

(b)

Figure 7.21 Equivalent circuit for p-i-n and APD detectors: (a) p-i-n; (b) APD

The bandwidth of the p-i-n detector is dictated by the transit time of the carrier across the i-layer and the circuit time constant established by C_o and any other external capacitance and the total resistive load.

The bandwidth due to the transit time effect is given by Equation 7.47 where w the depletion layer of the device, usually equal to the width of the i-layer, and v_{sat} the carrier velocity corresponding to the reverse bias. Usually this is the carrier saturation velocity.

$$\Delta f_{tr} = \frac{0.44}{w} v_{sat} \tag{7.47}$$

The bandwidth imposed by the circuit time constant is given by Equation 7.48 where C_t the total capacitance.

$$\Delta f_{CR} = \frac{1}{2 \pi R C_t} \tag{7.48}$$

C_t consists of the diode capacitance, the parasitic circuit capacitance and any other capacitance external to the detector. Therefore for high bandwidth all these capacitances must be as small as permitted by the technology used. Usually inequality 7.49 applies. However for very high bandwidth, that is in excess of 3GHz to 5GHz, the transit time starts limiting the device bandwidth and as a result the width of the i-layer is reduced with a subsequent reduction in n_q.

$$\Delta f_{tr} \gg \Delta f_{CR} \tag{7.49}$$

The noise of the p-i-n detector is due to the flow of current, shot noise, and due to thermal effects primary in R_s. The shot noise consists of two contributions; shot noise due to the dark current, given as usual by Equation 7.50, and shot noise due to the signal photocurrent itself, given by Equation 7.51 and the total p-i-n noise by 7.52.

$$< i_{dc}^2 > = 2 e I_{dc} \Delta f \tag{7.50}$$

$$< i_{sig}^2 > = 2\,e\,I_{ph}\,\Delta f \tag{7.51}$$

$$< i_{pin}^2 > = 2\,e\,(\,I_{ph}\,+\,I_{dc}\,)\,\Delta f \tag{7.52}$$

It should be clear now that even if the dark current is zero there will always be a noise component due to the signal. This noise-in-signal component establishes the performance limit of an optical receiver.

Because of the random nature of the impact ionisation process there is a fundamental difference between the noise of a p-i-n and that of an APD. The APD noise is given by Equation 7.53 where I_{dcm} is the part of the dark current subject to multiplication, I_{dcu} the un-multiplied dark current and F the excess noise factor.

$$< i_{apd}^2 > = 2\,e\,(\,I_{ph}\,+\,I_{dcm}\,)\,M^2\,F\,\Delta f + 2\,e\,I_{dcu}\,\Delta f \tag{7.53}$$

Had the ionisation process been deterministic then F = 1. In terms of gain, the excess noise factor can be expressed as in Equation 7.54.

$$F = M^x \tag{7.54}$$

The exponent x depends on the material and the design of the APD and for a well designed and used detector $0 < x < 1$.

The bandwidth of an APD depends on the circuit time constant but also on the gain-bandwidth product of the device, as given by Equation 7.55, where τ_{av} is the average transit time.

$$Gain - Bandwidth = \frac{3}{2\,\pi\,\tau_{av}} \tag{7.55}$$

The C_d of APDs is usually larger than that of p-i-n detectors because of the complexity of the devices. The state of the art of available p-i-n detectors is summarised in Table 7.4 and that of the APD in Table 7.5. Because of the superior performance of InGaAsP devices these tables do not include data on Ge detectors.

Table 7.4 Parameters for p-i-n detectors

Diameter (mm)	R (A/W)	I_d (nA)	C_d (pF)	Rise time (ns)	BW (MHz)	Material
0.24	0.8	0.5	1.5	1.0		Si
0.40	0.8	3.0	3.0	—	200.0	Si
0.08	0.8	5.0	1.0	—	1000	InGaAs
0.05	0.8	2.0	0.6	—	1000	InGaAs

Table 7.5 Parameters for avalanche photodetectors

Material	Diameter (μm)	R (A/W)	C_d (pF)	x	I_{dm} (nA)	BW (MHz)
Si	300	0.8	2.0	0.3	0.025	700
InGaAs	80	0.75	0.7	0.7	5.0	1200
InGaAs	50	0.75	0.4	0.7	4.0	2500

These detectors were selected as being suitable for communications. There are available even faster detectors but there are used mainly in optical measurements.

7.4 Optical receivers and transmitters

The function of the optical receiver is to recover the information imparted on one of the parameters of the optical carrier and in our case this means the intensity of the carrier. An optical receiver consists of an optical detector (the transducer) and a low noise electronic amplifier which raises the signal level to a value where further signal processing is possible without deterioration of the S/N ratio. After the low noise amplifier the details of the receiver depend on the application. An analogue receiver will include stages of amplification, filtering, electronic signal detection (whose function depends on the modulation format) and some additional signal processing such as filtering, and only after this is the information available to the user, as in Figure 7.22.

A digital receiver will include some more amplification, filtering, detection, clock recovery circuits, and finally the regenerator, after which the information is available to the user as in Figure 7.23. A combination is possible by using digital modulation on a subcarrier. In such a situation the subcarrier is first processed by an analogue receiver which recovers the digital information, but in analogue form,

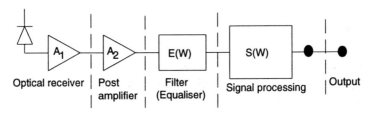

Figure 7.22   Analogue optical receiver

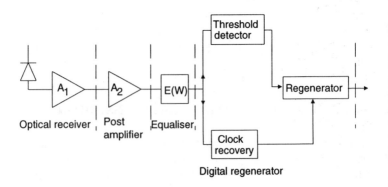

Figure 7.23   Digital optical receiver

and this is then processed by a digital receiver now minus the optical front end.

The measure of performance used in the design of optical receivers depends on the application. For analogue receivers the signal to noise ratio (SNR) is in widespread use. For digital receivers the bit error rate (BER) is used more or less universally.

The design of a complete receiver is a complex task and the purpose of this section is to concentrate on the design of the optical front end which consists of the detector and the low noise amplifier.

7.4.1 Receiver architecture

There are basically three classes of optical receivers: resistive load input, high input impedance or integrating receivers and low input impedance. The resistive load input optical receiver is the simplest possible, as shown in Figure 7.24. The detector is terminated on a resistor and the voltage developed across it is applied to the amplifier. The bandwidth is limited by the detector bandwidth, the bandwidth of the input circuit and the bandwidth of the amplifier itself. With a 50 ohm load the bandwidth can be very large. For example with a total capacitance of 1pF, a 50 ohms load and a 50 ohms amplifier input impedance the bandwidth is about 6.35GHz. Higher bandwidth can be obtained if the load is reduced and since commercially avail-

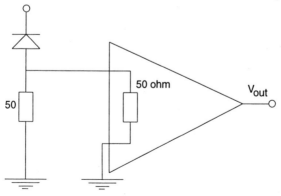

Figure 7.24   Resistive input optical receiver

able amplifiers have bandwidths of the order of 15GHz to 20GHz optical receivers with bandwidths of similar order are feasible. The difficulty with this approach lies in the high noise the resistive load input receiver yields compared to the other approaches. For this reason optical receivers designed for communication systems do not use the resistive load input configuration.

In a high input impedance or integrating receiver the photodetector is feeding directly a high to very high ohmic load, as in Figure 7.25. The signal developed across this resistor is fed to an active device of high input impedance such as a FET which acts as a voltage amplifier. After the first device there is subsequent amplification using amplifiers, with or without feedback according to requirements. The low frequency gain of the receiver is given by Equation 7.56 where A_{ampl} is the low frequency voltage gain of the amplifier.

$$A_{hi} = A_{ampl} R_L R P_{opt} \tag{7.56}$$

The receiver bandwidth is established by the first time constant at the input and it is given by Equation 7.57 where C_T is the total input capacitance.

$$f_1 = \frac{1}{2 \pi R_L C_T} \tag{7.57}$$

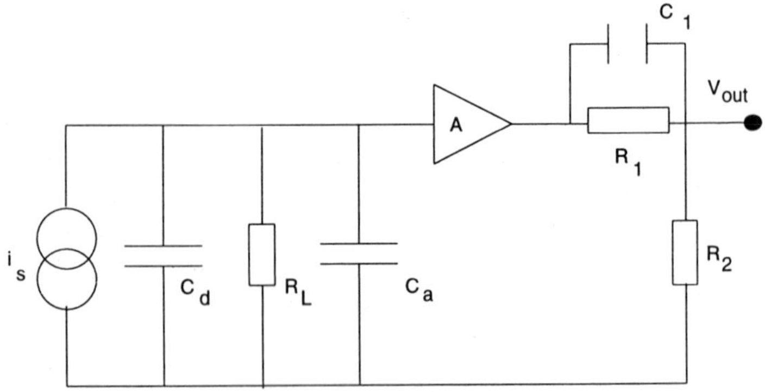

Figure 7.25 High input impedance optical receiver

The bandwidth is usually insufficient and some equalisation is required. The simplest approach is to cancel the pole by a zero using a RC network, as in Figure 7.25. The transfer function of this network is given by Equation 7.58 with the frequencies f_1 and f_2 given by Equations 7.59 and 7.60.

$$\frac{V_{out}}{V_{in}} = \frac{R_2}{R_1 + R_2} \frac{1 + j\dfrac{f}{f_1}}{1 + j\dfrac{f}{f_2}} \tag{7.58}$$

$$f_1 = \frac{1}{2 \pi R_1 C_1} \tag{7.59}$$

$$f_2 = \frac{R_1 + R_2}{2 \pi R_1 R_2 C_1} \tag{7.60}$$

The bandwidth of the integrating receivers is very small because the first pole is established by the high value load resistor and the input capacitance. This is of the order of 300kHz to 500kHz. However the bandwidth can be very large if the first pole is cancelled by a zero. With this approach bandwidths of the order of 10GHz to 15GHz are possible, provided the first pole is at a few MHz making the compensation easy. Integrating receivers with very large loads are used in low capacity systems such as 140MBit/s and below.

The low input impedance receivers use parallel input feedback at the point where the detector drives the amplifier, as in Figure 7.26, and they are also known as transimpedance receivers. In principle there is no need for high value resistive loads and the bandwidth of the receiver can be large. The receiver transfer function is given by Equation 7.61, with the bandwidth given by Equation 7.62.

$$\frac{v_o}{i_s} = \frac{- R_f}{1 + j 2 \pi R_f (C_f + C_T / A) f} \tag{7.61}$$

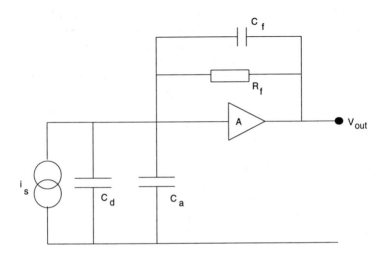

Figure 7.26 Transimpedance optical receiver

$$f_{BW} = \frac{A}{2\pi(C_T + A C_f)R_f} \tag{7.62}$$

Depending on the technology, bandwidths up to 5GHz are possible. Above that transimpedance receivers are not used because the use of feedback may lead to stability problems.

For larger bandwidths the receivers are designed with input impedances corresponding to that of the active devices, which are relative high but nowhere near the megaohm range used by truly integrating receivers. Equalisation is still necessary but the first pole is in the megahertz rather than the kilohertz range.

The main reason for using an integrating receiver is the capability of these receivers to deliver very low noise operation because the thermal noise is very small.

If APDs are available then the reason for using them is marginal. Their main disadvantage is that the very low bandwidth before equalisation requires the use of coding with small digital sum variation, (DSV).

As the bit rate increases coding becomes progressively difficult. The use of integrating receivers is expected to be further reduced with the introduction of signal formats with nearly uncontrollable DSV. Transimpedance receivers do not impose limits on the signal DSV and in spite of the fact that the noise performance with a p-i-n detector is inferior they dominate above 140MBit/s.

7.4.2 Analogue optical receivers

The SNR of an analogue receiver can be interpreted as either the SNR directly applicable to the information being received or as the carrier to noise ratio (CNR) which should then be used as the input SNR to the demodulator suitable for the modulation scheme employed. For example in a FM subcarrier modulation scheme the FM SNR can be obtain from the CNR by standard FM detection analysis.

The SNR or the CNR is given by Equation 7.63 in which m is the modulation index, $< i_{na}^2 >$ is the amplifier mean square noise and Δf is the receiver bandwidth.

$$SNR = \frac{1}{2} \frac{(m\, R\, M < P_{opt} >)^2}{[\, 2\, e\, (I_{dm} + R < P_{opt} >)\, M^2\, F + 2\, e\, I_{um}\,]\, \Delta f + < i_{na}^2 >}$$

$$(7.63)$$

The meaning of the other parameters can be found in the section on detectors. The details of the $< i_{na}^2 >$ depend on the representation of the amplifier noise.

For optical receivers the best approach is to represent the amplifier, being a single device or a multistage amplifier, by two independent noise sources with white spectral density, a current source i_n, and a voltage source e_n.

The detailed equations for these noise sources depends on the class of active devices used. For a FET amplifier they are given by Equations 7.64 and 7.65.

$$\frac{d < i_n^2 >_{fet}}{df} = 2\,e\,I_{gate} \tag{7.64}$$

$$\frac{d < e_n^2 >}{df} = \frac{4\,k\,T\,\Gamma}{g_m} \tag{7.65}$$

Γ is a numerical constant with a typical value of 0.7 for silicon FETs, 1.75 for GaAs MESFETs and 1.0 for short channel silicon MOSFETs. Then the noise spectral density referred to the input of the receiver is given by Equation 7.66 where R is either the load of the detector for high impedance input or the feedback resistor of a transimpedance amplifier.

$$\frac{d < n^2(f) >_{fet}}{df} = \frac{4\,k\,T}{R}$$

$$+ 2\,e\,I_{gate} + \frac{4\,k\,T\,\Gamma}{g_m}\,(2\,\pi\,C_t)^2 f^2 \tag{7.66}$$

C_t is the total input capacitance, which consists of the detector and input parasitic capacitance, the gate-source and gate-drain FET capacitances and the parasitic capacitance of the load resistor R.

The value of $< e_{na}^2 >$ is obtained by integrating Equation 7.66 over the bandwidth of the receiver. For a bipolar transistor amplifier the noise spectral density is given by Equation 7.67.

$$\frac{d < n^2(f) >_{bip}}{df} = \frac{4\,k\,T}{R} + 2\,e\,I_b + \frac{2\,e\,I_c}{g_m}\,(e\,\pi\,C_t)^2 f^2$$

$$+ 4\,k\,T\,r_{bb}\,(2\,\pi\,C_{sf})^2 f^2 \tag{7.67}$$

The capacitances C_T and C_{df} are given by Equations 7.68 and 7.69, where C_d is the parasitic detector capacitance and C_π and C_μ are the hybrid-π model capacitances.

$$C_T = C_{df} + C_\pi + C_\mu + C_f \tag{7.68}$$

$$C_{df} = C_{df} + C_f \tag{7.69}$$

7.4.3 Digital optical receivers

The design of a digital optical receiver is more complicated because first the performance index is the BER and secondly the receiver bandwidth depends both on the transmitted pulse and of receiver output pulse. The BER of a binary digital optical receiver is given by Equation 7.70, where the parameter Q is related to the signal to noise ratio.

$$BER = \frac{1}{\sqrt{2\pi}} \int_Q^\infty \exp\left(-\frac{x^2}{2}\right) dx \tag{7.70}$$

For Gausssina statistics and equal noise power for both ones and zeros, 2Q is the peak SNR required for a given BER. For large SNR a good approximation of the error integral is based on the asymptotic expansion, as in Equation 7.71 and for a BER of 10^{-9} Q=6.

$$BER = \frac{1}{\sqrt{2\pi}} \frac{\exp\left(-\frac{Q^2}{2}\right)}{Q} \tag{7.71}$$

Now in terms of Q the required optical power $< P_{opt} >$ for a p-i-n receiver is given by Equation 7.72.

$$< P_{opt} > = Q \frac{hf}{N_q e} (i_{na}^2)^{1/2} \tag{7.72}$$

The value of the reviver noise depends on the front end device. For a FET amplifier it is given by Equation 7.73, where I_2 and I_3 are the values of the Personic integrals that are depended on the shapes of the incident and out pulses and B the bit rate.

$$< is_{na}^2 > = 4 k \frac{T}{R} B I_2 + 2 e (I_d + I_g) B I_2$$

$$+ 4 k \frac{T}{g_m} (2 \pi C_T)^2 B^3 I_3 \tag{7.73}$$

With an incident non-return to zero (NRZ) pulse and a full raised cosine output pulse, $I_2 = 0.562$ and $I_3 = 0.0868$. With a return to zero (RZ) pulse, $I_2 = 0.403$ and $I_3 = 0.0984$. It is clear that since the optical power depends on the cube of the bit rate the receiver design should minimise the C_T. With a bipolar front end receiver Equation 7.74 is obtained.

$$< i_{na}^2 >_{bip} = 4 k \frac{T}{R} B I_2 + 2 e I_b B I_2$$

$$+ 2 e \frac{I_c}{g_m} (2 \pi C_t)^2 B^3 I_3 + 4 k T r_{bb} 2 \pi C_{sf} \tag{7.74}$$

The receiver sensitivity equation for an APD detector is given by Equation 7.75, where the un-multiplied dark current was assumed to have negligible effect on the sensitivity and I_1 the value of another Personick integral which equals 0.5 for both NRZ and RZ pulse formats.

$$< P_{opt} > = Q \frac{h f}{n_q e} \left[\left(\frac{< i_{na}^2 >}{M^2} + 2 I_{dm} F B I_2 \right)^{1/2} \right.$$

$$\left. + e F B I_1 Q \right] \tag{7.75}$$

The details of the design of optical receivers depend on the applications but as a general guideline low noise operation is achieved by minimising the input capacitance of the receiver.

Because of the importance of the optical receivers, there has been a large number of designs aiming at particular requirements. How-

Table 7.6 Sensitivity of optical receivers using p-i-n detectors. (All parameters are at BER = 10^{-9}; all the receivers are with multimode fibre tails)

Bit rate	Sensitivity (dBm)	
	Min	Typ
2Mbit/s	−56.5	−58.5
16Mbit/s	−52.0	−54.0
45Mbit/s	−49.0	−51.0
90Mbit/s	−46.0	−48.0
160Mbit/s	−43.0	−45.5
565Mbit/s	−36.0	−38.0
680Mbit/s	−34.0	−35.5
1200Mbit/s	−31.0	−33.0
1600Mbit/s	−29.0	−31.0
2400Mbit/s	−28.0	−29.0

ever, since the main use of optical receivers is in the trunk and junction network, where the information capacity is standardised, receivers corresponding to these rates have been introduced. Table 7.6 summarises the current performance of optical receivers using p-i-n detectors and operating in the 1300nm to 1600nm range. APD receivers tend to be in house products and the improvement depends on the details of the design but on average 5db of improvement is expected.

Some of the bit rates are not within the digital hierarchy, but in a fast evolving field some of the applications require special bit rates.

7.4.4 Optical transmitters

The configuration of an optical transmitter depends first on the signal format and second on the class of the device (LED or laser).

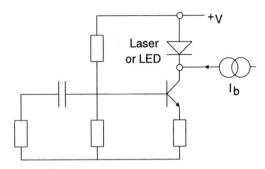

Figure 7.27 Driver for analogue modulation

The simplest modulation format is that of intensity modulation. In this format the information carrying current is injected into the optical source and consequence the intensity, that is the square of the field, is modulated.

For analogue modulation the optical source is biased at a given point and the modulation current is superimposed on it. Figure 7.27 shows a driver which can be used for analogue modulation.

The modulation waveforms vary depending on whether the optical device is a LED or a laser, as shown in Figures 7.28 and 7.29. Clearly there are issues of linearity as with any analogue system which has to be resolved. The usual techniques of feedback and pre-distortion can be used but improved device performance is the critical issue in analogue modulation.

For digital modulation the device is based either around zero (LEDs) or at threshold (lasers). If lasers are biased at zero current then there is significant turn on delay, which impairs the performance at high bit rates. The typical high speed digital modulator is current mode logic otherwise known as the emitter coupled switch, shown in Figure 7.30. The advantage of this configuration is the large bandwidth available and the ability to control the modulation current through the constant current source at the tail of the switch.

In addition to the actual modulator an optical transmitter is required to maintain a constant output optical power. This is far more important for lasers which are very sensitive to temperature. A typical

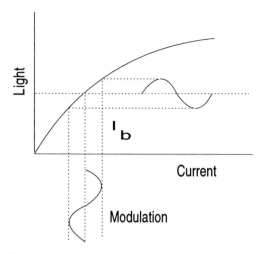

Figure 7.28 Analogue LED modulation

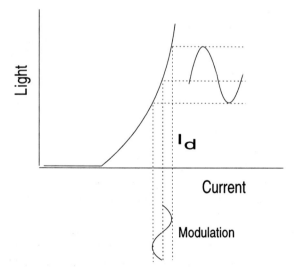

Figure 7.29 Analogue laser modulation

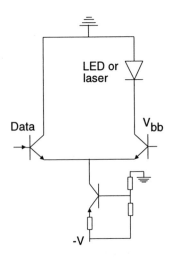

Figure 7.30 Digital modulator for LEDs and lasers

arrangement to maintain the output power constant is shown in Figure 7.31.

The photodiode monitors the power at the back facet of the laser and produces a signal proportional to the average optical power. This signal is then compared with a reference and the controller adjusts the laser bias current accordingly. This control scheme cannot control the level of 'ones' and 'zeros' individually and for some advanced applications, such as long haul systems, more elaborate schemes are used which control the level of 'ones' and 'zeros'.

7.5 Optical system design

The design of optical systems follows the same approach as any other communication system. The approach can be summarised in a number of steps as follows:

1. The required performance against an agreed performance index is established.
2. The basic system parameters are selected.

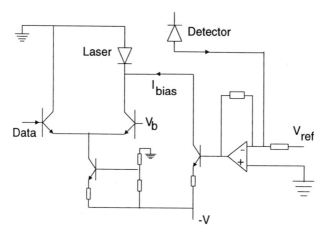

Figure 7.31   Mean optical power controller

3.   The performance of suitable subsystems is established.
4.   Allocations are made for the implementation penalties of subsystems.
5.   The impact of environmental factors is assessed in terms of performance impairment.
6.   The system operational margins are introduced.
7.   On the basis of (3) to (6) the expected performance is estimated and compared to the required one from (1).
8.   If the requirements are not satisfied then a design iteration begins, covering steps (2) to (7).

The design therefore is in both principle and practice an iterative process and some of the parameters entering the iteration depend on existing practices and understanding rather than scientific principles. This particularly applies to steps (4) and (6) above.

A methodology of system design as summarised in steps (1) to (8) can be treated completely in the abstract in terms of system theory, but it will be far easier to describe system design by analysing a concrete case. Since the approach applies to both analogue and digital systems only digital system shall be discussed. However, before system design examples are discussed it is important to examine the question of coding in optical fibre systems.

7.5.1 Coding in optical fibre systems

The coding of the transmitted information has been used in coaxial systems to reduce the bandwidth and make it possible to increase the section length and as a result minimise the number of repeaters in a system. This was at the expense of increased terminal complexity, but overall it yielded cost effective systems. With the advent of fibre communications the issue of bandwidth lost its importance because the bandwidth of single mode fibre is very large. For this reason, in the initial stages of the introduction of fibre communications, coding was used to ensure that the transmitted signal had sufficient clock content so that low Q clock extraction circuits could be used. In principle of course that was not necessary.

The requirement for a code to perform a function of primary importance emerged with the use of integrating optical receivers. Since the dynamic range of an integrating receiver is limited there is the need for codes with small digital sum variation (DSV). Such codes turn out to be invariably block codes. In a block code a number of information digits, say N, are mapped through deterministic rules into M output digits with M > N. This will lead to a slight increase of the bit rate in the fibre and in error extension, but the small penalty in receiver sensitivity is worth paying. Block codes such as 5B6B and 7B8B have been used in optical systems up to 565MBit/s.

As the information capacity increased it became progressively more difficult to design such coding schemes because of the limitations in the speed of the electronics. In addition the requirements for information transparency forced the design of optical receivers capable of handling signal formats with very high DSV. These receivers offer sufficient sensitivity to make possible cost effective systems even if their sensitivity is not as good as that of the integrating receivers. Taking into consideration all these factors, state of the art fibre systems used codes of very low redundancy for data management purposes. These low redundancy data sequences are scrambled to minimise the requirements on the clock extraction subsystem in the receiver. Block codes are not error correction codes and in principle error correction can be used in optical systems especially where an error floor is introduced. Again, the complexity and speed requirements of the electronics are such that error correction at the fibre bit rate is hardly used. Error correction is used at the low capacity tributary channels where, due to the relative low speed, high levels of integration are possible.

7.5.2 Design of multimode systems

Consider as the design target the maximisation of the section length of a digital system operating at 140Mbit/s and using a directly modulated laser with 1mW (OdB) output power and a spectral width under modulation of 3nm.

Since we want to maximise the length the obvious choice of wavelength is the 1550nm window. At 1550nm the fibre loss is around 0.3dB/km with an intermodal dispersion of 120ps/km and a intramodal dispersion of 17ps/nm.km. With a 3nm spectral width the total dispersion is 130ps/km. Assuming that 1dB dispersion penalty can be tolerated the maximum dispersion limited section length for a quarter of a bit duration dispersion is 13.7km. This length corresponds to a fibre loss of 4.1dB and the system is dispersion limited. Because of that one can design a receiver with reduced sensitivity using a p-i-n detector. For the example discussed here let us assume that the theoretical receiver sensitivity is –30dBm. The various impairments and margins introduced reflect current practice and they are expected to change as better understanding is achieved. The system power budget is summarised in Table 7.7.

The additional system margin can be used in a variety of ways. For example if the dispersion penalty increases to 3dB the dispersion limited section length increases to 19.2km. This represents an additional total loss of 3.65dB reducing the margin to 5.25dB.

This example indicates the limitations of multimode transmission for high bit rates and low loss fibre. Had the dispersion been negligible the section loss capability would have been equivalent to 46.6km and this big difference is the fundamental reason for the introduction of single mode transmission.

Table 7.7 System power budget

Transmitter output (dBm)		−3.0
Receiver sensitivity (dBm)		−30.0
Available section loss capability (dBm)		27.0
Transmitter penalties		
Implementation penalties (dB)	0.5	
Temperature effects (dB)	1.0	
Ageing effects (dB)	0.5	
Total transmitter penalty (dB)		2.5
Receiver penalties		
Implementation (dB)	1.5	
Temperature effects (dB)	1.0	
Digital regenerator (dB)	1.5	
Ageing (dB)	0.5	
Total receiver penalty (dB)		4.5
System margins		
Operating margin (dB)	3.0	
Connectors (pair) (dB)	1.0	
Cable repairs (dB)	2.0	
Dispersion penalty (dB)	1.0	
Total system margin (dB)		7.0
Section loss capability (dB)		13
Section loss (dB)		4.1
Additional system margin (dB)		8.9

7.5.3 Design of single mode systems

As a design example for a single mode system consider the problem of maximising the section length of a 2.4Gbit/s system. Again the optimum wavelength window is around 1550nm where top quality single mode fibre has a loss of 0.2dB/km. The dispersion at 1550nm depends on the class of the fibre. Fibre designed for zero dispersion at 1300nm has 16 to 18ps/km.nm dispersion at 1550nm. On the other hand fibre designed for zero dispersion at 1550nm is usually specified as having a window within which the dispersion is less than a given number. A usual number for this tolerance is 3ps/km.nm.

In the design of high capacity single mode systems using non-dispersion shifted fibre one of the problems is laser chirping. Chirping takes place because in directly modulated lasers the injection of the modulation current forces a small but nevertheless crucial wavelength shift in the single longitudinal mode of the device. For well designed DFB lasers the chirping is of the order of 0.1nm to 0.3nm.

In spite this small value chirping imposes a significant limitation on the section length. For example with conventional fibre of 16ps/km.nm dispersion and a 0.2nm chirp the 1db dispersion penalty corresponds to 32km. Again the system is dispersion limited but this time not exclusively due to dispersion but also to the interaction of laser dynamics (chirp) with dispersion. With a DSF the 1dB dispersion penalty corresponds to 173km.

The dispersion penalty for $(LD(\lambda)\Delta\lambda <) t_c$, where t_c is half the period of the laser relaxation oscillations, is given by Equation 7.76, and for $LD(\lambda)\Delta\lambda > t_c$ by Equation 7.77.

$$Penalty\ (dB)\ =\ 10\log\left(\frac{1}{1-4\,LD(\lambda)\,B\,\Delta\lambda}\right) \tag{7.76}$$

$$Penalty\ (dB)\ =\ 10\log\left(\frac{1}{1-4\,B\,t_c}\right) \tag{7.77}$$

For example assume that $t_c = 80$ps. Then with $\Delta\lambda = 0.15$nm the maximum $LD(\lambda)$ product is 0.533ns/nm. For $LD(\lambda) - 0.5$ the penalty is 5.53 at 2.4Gbit/s.

The obvious solution to the chirp problem with directly modulated lasers is to use DSF. However, the use of external modulators such as LiNbO or electro-optic absorption devices offers possible alternatives if the use of DSF is not possible. This is particularly important in upgrading installed NDSF fibre. The design of a single mode system is similar to that of a multimode system outlined earlier.

7.6 Fibre optic applications

The basic characteristics of fibre optic communications are the low loss and large bandwidth of the channel (the fibre), the high performance, compactness and reliability of the components (sources and detectors) and the high performance subsystems possible (optical receivers and transmitters). These features, combined with the rapid progress made in integrated electronics, has ensured fibre optic penetration into virtually all the communication applications ranging from submarine systems to data buses for avionics. Considering that it was only twenty years ago when it became clear that it was possible to realise the potential of fibre transmission this rapid and virtually complete acceptance of the new technology is quite unusual. The impact of fibre optic communications lies not so much in that they can do better than other technologies but they have altered the way the communication issues of the next century will be approached.

7.6.1 Submarine fibre optic systems

The impact of fibre communications in submarine communications has been greater than that on terrestrial applications. In submarine applications not only the medium has changed, from coaxial cable to fibre, but also the transmission format, the system configuration and the long term prospects of submarine communications. The key to this change is the high reliability of the electro-optics components. Submarine systems are planned and designed for 25 year service with a maximum of three repairs, because repairs of these systems are time consuming and expensive operations.

The first submarine system was the Transatlantic telephone cable (TAT-8) connecting the USA with the UK and France. The distance is about 6000km and the information capacity of the system was 560Mbit/s realised with 2 × 80Mbit/s channels. The loss and dispersion requirements were satisfied by operating single mode at 1300nm with a chromatic dispersion of less that 2ps/km.nm, a Fabry-Perot laser and a InGaAsP p-i-n detector followed by an integrated Si transimpedance receiver. The receiver sensitivity was −31dBm, − 2dBm optical power and with 6dB total margin the available section loss was 23dB including the splice losses. With 0.45dB/km loss the section length was 50km.

The system was commissioned in 1988 and its success encouraged the installation of other systems such as the Trans Pacific Cable TPC-3 (2 × 280Mbit/s; 1988) the California to Hawaii HAW-1 (2 × 280 Mbit/s; 1988), the Private Transatlantic Cable PTAP-1 (2 × 420 Mbit/s; 1989), the North Pacific Cable NPC (3 × 565 Mbit/s; 1990) and the TAT-9 system (2 × 560 Mbit/s; 1991). TAT-9 is the most complex of the submarine systems designed. At the American end there are two spurs. One from Canada (220 km) and one from the USA (1320 km). They combine in an undersea branching multiplexer/demultiplexer. Then there are 4600km across the North Atlantic to another undersea multiplexer/demultiplexer out of which there are three spurs. One to UK (530km), one to France (302km) and one to Spain (1390km). This system is the first to use the 1550nm window with DFB lasers, p-i-n detectors and integrated silicon bipolar receivers. The section length is 120km using NDSF.

7.6.2 Terrestrial fibre optic systems

The first application of fibre optic communications were in terrestrial systems. The first systems were operating at 850nm using multimode transmission but the advances made in the performance of the fibre at the 1300nm window and of the electro-optics forced the migration of the applications to this wavelength.

Bit rates of 140MBit/s and above are used in the trunk network and those below in the junction network. Trunk systems use lasers and high sensitivity p-i-n/FET or APD receivers and because the aim is to maximise section length, single mode fibres are used. Junction systems usually use LEDs and multimode fibre because the dispersion penalty is negligible.

The length of trunk system varies and depends on the distribution of population. System length can be as short as 50km but as long as 1000km. In Europe and some areas of the USA system lengths are of the order of 100km to 150km. The length of junction systems is around 2km to 25km. Therefore the benefits of using fibre in long haul transmission are not apparent in the junction network. Nevertheless, fibre optic systems are used in the junction network because they are able to operate without repeaters in the congested environment of the urban areas.

The majority of optical systems currently in operation use the 1300nm window but the requirements for higher capacity, that is 2.4Gbit/s and higher, and the trend for un-repeated transmission ensured that future systems will operate at 1550nm. The technology needed for this migration is DFB lasers and long wavelength APDs. They are both beginning to be available in volume and at prices which will lead to cost effective systems.

7.7 The future of fibre networks

In the early 1980's an attempt was made in North America to address the shortcomings of the asynchronous multiplexing structure then in use. The main task was to try and avoid the multiplexed signal 'hiding' its contents amongst control and stuffing bits. The only way to extract information in this format was to completely demultiplex the signal to extract a single channel. A standard called Syntran was the result of this work supported primarily by Bellcore (Bell Communications Research). Syntran, however, as its name suggests, required a synchronous environment in order to permit access to any low order signal within the multiplexed high order signal without having to demultiplex the whole thing.

Needing a synchronous network is not a bad idea, in fact it is the whole basis of the new Synchronous Digital Hierarchy transmission standards (known as SONET in North America), however, at the time Syntran was introduced the benefits of using it did not overcome the difficulties of implementing it. Consequently, Syntran saw very little application in the North American network.

At about the same time that Syntran was being developed there was an emerging need in the fibre optic network to develop mid span meet standards. That is, a standard that would allow different manufacturers products on each end of a fibre to communicate. At the time each vendor used proprietary methods of encoding information into an optical signal; the only common interface being at the electrical asynchronous ports.

As a result of this need for easy access to low order signals and mid span meet on fibres the new SONET (Synchronous Optical Network) standard was developed in North America. Before the standard was completed there was significant interest expressed by the ITU-T standards body, resulting in a modification to SONET to make it adaptable to ITU-T plesiochronous bit rates. The standard issued by ITU-T became known as SDH (Synchronous Digital Hierarchy).

7.7.1 Basics of SONET/SDH

The principle behind the standard is to create a synchronous fibre optic network that can accept asynchronous electrical tributary signals and carry them through the network in payloads, the payload being of a higher bandwidth than the tributary. The tributary is allowed to 'float' around within the payload due to the difference in clock rates between the synchronous fibre network and the asynchronous electrical network. As the tributary floats around it is necessary to know exactly where it is should it need to be extracted for termination at its destination in the network. This is done through the use of pointer tables that are attached to a payload as overhead bytes. The pointer table identifies where within the payload the first byte of the tributary is located.

The lowest rate SDH signal is called STM1 (Synchronous Transport Module) and is defined at 155.52mbit/s, including payload and overhead. Higher rate signals are exact multiples of the STM1 signal. Each of these signals has the same structure: floating payloads of tributary traffic and an overhead carrying pointers. The overhead actually carries other information besides the pointers. It carries information channels that allow the SDH products to communicate with each other throughout the network. They can exchange maintenance information (alarms, error rates, protection switching etc.) that permits widespread network management and control from a single point.

Compared to plesiochronous (or asynchronous in North America), SDH brings five major benefits to the network that make deployment very worthwhile:

1. Multi-vendor. There is now a standard allowing for the mixing of products from a variety of vendors without having to be concerned about interfaces. In fact this history of interface issues forced the structure of the telephone network into categories of switching and transmission with well defined boundaries at which interface problems were dealt with. This need no longer be true.

2. Flexibility. As SDH now allows us to reach in to any multiplexed signal and extract only the information we need it frees the network designer from a cost problem. The problem used to be that back to back multiplexers had to be used to perform this add/drop function which was very costly. As a result this tended to be avoided or centralised to reduce costs, either of which led to difficult engineering rules that limited service responsiveness to the end user.

3. Control. Combined with the flexibility inherent in the SDH network, the control available through the overhead maintenance channels gives the telephone company the ability to respond to customer service needs immediately. The software intensive design of these new generation products also means that the telephone company will be able to dialogue with individual terminals and obtain status reports on performance.

4. Reliability. SDH will have an overall effect of simplifying the network. An SDH terminal performs several levels of multiplexing and optical conversion all in one unit thus reducing the number of network elements required. A simpler network combined with sophisticated control capability and the intel-

ligence in each terminal to report on its health (and correct it if necessary) results in a very reliable network.

5. Transmission capacity. Because SDH is a standard for fibre transmission it brings all the benefits of the bandwidth of fibre to the network. An STM16 product has a 2.4Gbit/s capacity, far greater than todays plesiochronous products. And this is not the limit since the standard could support up to 9.6Gbit/s on conventional single mode fibre cable! Now the network need not be service restricted due to bandwidth bottlenecks.

A typical SDH based network is shown in Figure 7.32.

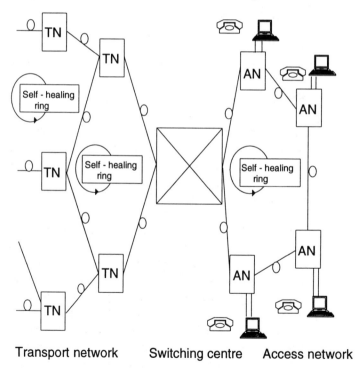

Figure 7.32 Typical end to end Sonet/SDH network with integral OAM&P

The fibre optic cable can be directly connected to the switch, since all of the SDH based bandwidth can now be managed by software control. As fibres enter the central office there would be no need for patch panels or cross-connects as all bandwidth assignment and capacity grooming will be performed throughout the network at each network element. Nor will it be necessary to segregate traffic in the central office for management based on type of service. All fibres will connect to one switch capable of handling all services whether analog or digital; voice, data or video; narrowband, wideband or broadband; connection oriented or connectionless.

Traffic needing to go on into the network will do so on very high capacity fibres again configured as rings for survivability. Service provisioning now becomes a keyboard entry that is interpreted and transmitted to all network elements immediately, providing rapid response to the end user.

7.8 References

Alferness, R.C. (1995) Advanced technologies pave the way for photonic switches, *Laser Focus World*, February.

Black, A. (1994) Achieving the best possible junction, *Cabling World*, December.

Boettle, et al. (1994) Towards all-optical networking, *Electrical Communication*, 3rd Quarter.

Carroll, C. (1994) The fiber versus copper decision, *Lightwave*, January.

Chapman, D.A. (1994) Erbium-doped fibre amplifiers: the latest revolution in optical communications, *Electronics & Communication Engineering Journal*, April.

Chesnoy, J. et al. (1994) Ultrahigh bit rate transmission for the years 2000, *Electrical Communication*, 3rd Quarter.

Garrett, K. (1995) Optical switches benefit ATM network capacities, efficiencies and costs, *Lightwave*, February.

Higgins, R. (1994) Understanding fibre cabling, *Cabling World*, December.

Higins, T.V. (1994) Creating laser light, *Laser Focus World*, June.

Kessler, J.N. (1994) Fibre optics: a 10 year view, *Lightwave*, January.

Kolodziej, S. (1994) Fibre optic cables overcome Category 5 copper wire problems in local networks, *Lightwave*, November.

Lacroix, Y. and Vella, P. (1995) How to optimise the performance of fibreoptic amplifiers, *Laser Focus World*, January.

Muoi, T.V. (1994) Fibreoptic transmitters and receivers meet system needs, *Laser Focus World*, June.

Nelson, W.H. et al. (1994) Optical switching expands communication network capacity, *Laser Focus World*, June.

Offside, M.J. et al. (1995) Optical wavelength converters, *Electronics & Communication Engineering Journal*, April.

Pan, J.J. and Liang, F. (1994) Optoelectronic components make WDM networks practical, *Laser Focus World*, January.

Pitassi, et al. (1994) Fusion splicing: a history, *Lightwave*, October.

Rehm, W. et al. (1995) Optical interconnects: the key to high bitrate communication within telecommunication equipment, *Electrical Communication*, 1st Quarter.

Snell, M. (1995) Travel light, travel fast, *Communications International*, February.

Uttamchandani, D. (1994) Fibre-optic sensors and smart structures: devvelopments and prospects, *Electronics & Communication Engineering Journal*, October.

8. Acronyms

Every discipline has its own 'language' and this is especially true of telecommunications, where acronyms abound. In this guide to acronyms, when the letters within an acronym can have slightly different interpretations, these are given within the same entry. If the acronym stands for completely different terms then these are listed separately.

ABM	Asynchronous Balanced Mode.
ACK	Acknowledgement. (Control code sent from a receiver to a transmitter to acknowledge the receipt of a transmission.)
ADC	Analogue to Digital Conversion.
ADM	Adaptive Delta Modulation. (Digital signal modulation technique.)
ADM	Add-Drop multiplexer. (Term sometimes used to describe a drop and insert multiplexer.)
ADP	Automatic Data Processing.
ADPCM	Adaptive Differential Pulse Code Modulation. (ITU-T standard for the conversion and transmission of analogue signals at 32kbit/s.)
AE	Anomaly Events. (E.g. frame errors, parity errors, etc. ITU-T M.550 for digital circuit testing.)
AM	Amplitude Modulation. (Analogue signal transmission encoding technique.)
AMI	Alternate Mark Inversion. (Line code system.)
ANBFM	Adaptive Narrow Band Frequency Modulation.
ANBS	American National Bureau of Standards.
ANSI	American National Standards Institute.
APC	Adaptive Predictive Coding.
APD	Avalanche Photodiode. (A semiconductor device used for fibre optic communications.)

APK   Amplitude Phase Keying. (A digital modulation technique in which the amplitude and phase of the carrier are varied.)

ARQ   Automatic Request for repetition. (A feature in transmission systems in which the receiver automatically asks the sender to retransmit a block of information, usually because there is an error in the earlier transmission.)

ASCII   American Standard Code for Information Interchange. (Popular character code used for data communications and processing. Consists of seven bits, or eight bits with a parity bit added.)

ASK   Amplitude Shift Keying. (Digital modulation technique.)

ATDM   Asynchronous Time Division Multiplexing.

AVDM   Analogue Variable Delta Modulation.

BCC   Block Check Character. (A control character which is added to a block of transmitted data, used in checking for errors.)

BCD   Binary Coded Decimal. (An older character code set, in which numbers are represented by a four bit sequence.)

BCH   Bose Chaudhure Hocquengherm. (Coding technique.)

BELLCORE   Bell Communications Research. (Research organisation, incorporating parts of the former Bell Laboratories, established after the divestiture of AT&T. Funded by the BOCs and RBOCs to formulate telecommunication standards.)

BER   Bit Error Ratio. (Also called Bit Error Rate. It is a measure of transmission quality. It is the number of bits received in error during a transmission, divided by the total number of bits transmitted in a specific interval.)

BERT   Bit Error Ratio Tester. (Equipment used for digital transmission testing.)

BIP	Bit Interleaved Parity. (A simple method of parity checking.)
BIST	Built In Self Test.
BISYNC	Binary Synchronous communications. (Older protocol used for character oriented transmission on half-duplex links.)
BnZS	Bipolar with n-Zero Substitution. (A channel code. Examples are B3ZS which has three-zero substitution; B6ZS with six-zero substitution, etc.)
BPSK	Binary Phase Shift Keying.
BPV	Bipolar Violation. (Impairment of digital transmission system, using bipolar coding, where two pulses occur consecutively with the same polarity.)
BRZ	Bipolar Return to Zero. (A channel coding technique, used for digital transmission.)
BSI	British Standards Institute.
CCITT	Comite Consultatif Internationale de Telephonique et Telegraphique. (Consulative Committee for International Telephone and Telegraphy. Standards making body within the ITU, now forming the new Standardisation Sector and referred to as ITU-T.)
CDM	Code Delta Modulation. (Or Continuous Delta Modulation.)
CFM	Companded Frequency Modulation.
CFSK	Coherent Frequency Shift Keying.
CMT	Character Mode Terminal. (E.g. VT100, which does not provide graphical capability.)
CODEC	COder-DECoder.
CPFSK	Continuous Phase Frequency Shift Keying.
CPSK	Coherent Phase Shift Keying.
CPU	Central Processing Unit. (Usually part of a computer.)
CRC	Cyclic Redundancy Check. (Bit oriented protocol used for checking for errors in transmitted data.)
CRT	Clear To Send. (Control code used for data transmission in modems.)

CVSD	Continuous Variable Slope Delta modulation. (Proprietary method used for speech compression. Also called CVSDM.)
CW	Continuous Wave.
DBT	Deutsche Bunderspost Telekom. (Also written DBP-T. German PTT.)
DCA	Dynamic Channel Allocation. (System in which the operating frequency is selected by the equipment at time of use, rather than by a planned assignment.)
DCC	Data Communication Channel.
DCDM	Digitally Coded Delta Modulation. (Delta modulation technique in which the step size is controlled by the bit sequence produced by the sampling and quantisation.)
DCE	Data Circuit termination Equipment. (Exchange end of a network, connecting to a DTE. Usually used in packet switched networks.)
DCF	Data Communications Function.
DCN	Data Communications Network.
DE	Defect Events. (E.g. loss of signal, loss of frame synchronisation, etc. ITU-T M.550 for digital circuit testing.)
DEDM	Dolby Enhanced Delta Modulation.
DEPSK	Differentially Encoded Phase Shift Keying.
DM	Degraded Minutes. (Any one minute period with a BER exceeding 10^{-6}, as per ITU-T G.821.)
DM	Delta Modulation. (Digital signal modulation technique.)
DMA	Direct Memory Access.
DPCM	Differential Pulse Code Modulation.
DQPSK	Differential Quaternary Phase Shift Keying.
DS-0	Digital Signal level 0. (Part of the US transmission hierarchy, transmitting at 64kbit/s. DS-1 transmits at 1.544Mbit/s, DS-2 at 6.312Mbit/s, etc.)
DSAP	Destination Service Access Point. (Refers to the address of service at destination.)

DSB	Double Sideband.
DSBEC	Double Sideband Emitted Carrier.
DSBSC	Double Sideband Suppressed Carrier modulation. (A method for amplitude modulation of a signal.)
DSM	Delta Sigma Modulation. (Digital signal modulation technique.)
DSP	Digital Signal Processing.
DTE	Data Terminal Equipment. (User end of network which connects to a DCE. Usually used in packet switched networks.)
DTI	Department of Trade and Industry.
EBCDIC	Extended Binary Coded Decimal Interchange Code. (Eight bit character code set.)
ECC	Error Control Coding. (Coding used to reduce errors in transmission.)
EFS	Error Free Seconds. (In transmitted data it determines the proportion of one second intervals, over a given period, when the data is error free.)
EHF	Extremely High Frequency. (Usually used to describe the portion of the electromagnetic spectrum in the range 30GHz to 300GHz.)
EOA	End Of Address. (Header code used in a transmitted frame.)
EOB	End Of Block. (Character used at end of a transmitted frame. Also referred to as End of Transmitted Block or ETB.)
EOC	Embedded Operations Channel. (Bits carried in a transmission frame which contain auxiliary information such as for maintenance and supervisory. This is also called a Facilities Data Link, FDL.)
EOT	End Of Transmission. (Control code used in transmission to signal the receiver that all the information has been sent.)
ERL	Echo Return Loss.
ESF	Extended Superframe. (North American 24 frame digital transmission format.)

ETB	End of Transmission Block. (A control character which denotes the end of a block of Bisync transmitted data.)
ETX	End of Text. (A control character used to denote the end of transmitted text, which was started by a STX character.)
FAS	Frame Alignment Signal. (Used in the alignment of digital transmission frames.)
FCS	Frame Check Sequence. (Field added to a transmitted frame to check for errors.)
FDM	Frequency Division Multiplexing. (Signal multiplexing technique.)
FDMA	Frequency Division Multiple Access. (Multiple access technique based on FDM.)
FDX	Full Duplex. (Transmission system in which the two stations connected by a link can transmit and receive simultaneously.)
FEC	Feedforward Error Correction. (Also called Forward Error Correction. Technique for correcting errors due to transmission.)
FEXT	Far End Crosstalk.
FFSK	Fast Frequency Shift Keying.
FM	Frequency Modulation. (Analogue signal modualtion technique.)
FMFB	Frequency Modulation Feedback.
FSK	Frequency Shift Keying. (Digital modulation.)
GHz	Giga Hertz. (Measure of frequency. Equal to 1000000000 cycles per second. See Hertz or Hz.)
GMSK	Gaussian Minimum Shift Keying. (Modulation technique, as used in GSM.)
GoS	Grade of Service. (Measure of service performance as perceived by the user.)
HDB3	High Density Bipolar 3. (Line transmission encoding technique.)

HDLC	Higher level Data Link Control. (ITU-T bit oriented protocol for handling data.)
HOMUX	Higher Order Multiplexer.
HRDS	Hypothetical Reference Digital Section. (ITU-T G.921 for digital circuit measurements.)
Hz	Hertz. (Measure of frequency. One Hertz is equal to a frequency of one cycle per second.)
IA2	International Alphabet 2. (Code used in a tele-printer, also called the Murray code.)
IA5	International Alphabet 5. (International standard alphanumeric code, which has facility for national options. The US version is ASCII.)
IDN	Integrated Digital Network. (Usually refers to the digital public network which uses digital transmission and switching.)
IF	Intermediate Frequency.
IFRB	International Frequency Registration Board. (Part of the ITU's Radiocommunication Sector.)
IM	Intermodulation.
I/O	Input/Output. (Usually refers to the input and output ports of an equipment, such as a computer.)
IR	Infrared.
IRED	Infrared Emitting Diode.
ISI	Inter-Symbol Interference. (Interference between adjacent pulses of a transmitted code.)
ISM	Industrial, Scientific and Medical. (Usually refers to ISM equipment or applications.)
ISO	International Standardisation Organization.
IT	Information Technology. (Generally refers to industries using computers e.g. data processing.)
ITA	International Telegraph Alphabet.
ITU	International Telecommunication Union.
KBS	Knowledge Based System.
kHz	KiloHertz. (Measure of frequency. Equals to 1000 cycles per second.)

LASER	Light Amplification by Stimulated Emission of Radiation. (Laser is also used to refer to a component.)
LDM	Linear Delta Modulation. (Delta modulation technique in which a series of linear segments of constant slope provides the input time function.)
LED	Light Emitting Diode. (Component which converts electrical energy into light.)
LISN	Line Impedance Stabilising Network. (An artificial network used in measurement systems to define the impedance of the mains supply.)
LJU	Line Jack Unit.
LPC	Linear Predictive Coding. (Encoding technique used in pulse code modulation.)
LRC	Longitudinal Redundanc Check. (Error checking procedure for transmitted data.)
LSB	Least Significant Bit. (Referring to bits in a data word.)
LTE	Line Terminating Equipment. (Also called Line Terminal Equipment. Equipment which terminates a transmission line.)
MAC	Media Access Control. (IEEE standard 802. for access to LANs.)
MHz	MegaHertz. (Measure of frequency. Equal to one million cycles per second.)
MIPS	Million Instructions Per Second. (Measure of a computer's processing speed.)
MODEM	MOdulator/DEModulator. Device for enabling digital data to be send over analogue lines.
MOS	Metal Oxide Semiconductor. (A semiconductor technology.)
MOSFET	Metal Oxide Semiconductor Field Effect Transistor. (A transistor made from MOS.)
MOTIS	Message Oriented Text Interchange System. (ISO equivalent of a message handling system.)
MoU	Memorandum of Understanding.

MSK	Minimum Shift Keying. (A form of frequency shift keying, or FSK.)
MTBF	Mean Time Between Failure. (Measure of equipment reliability. Time for which an equipment is likely to operate before failure.)
MUF	Maximum Usable Frequency.
NA	Numerical Aperture. (Measure of a basic parameter for fibre optic cables.)
NAK	Negative Acknowledgement. (In data transmission this is the message sent by the receiver to the sender to indicate that the previous message contained an error, and requesting a re-send.)
NEXT	Near End crosstalk. The unwanted transfer of signal energy from one link to another, often closely located, at the end of the cable where the transmitted is located.
NF	Noise Figure.
NFAS	Not Frame Alignment Signal. (In transmitted code.)
NNI	Network Node Interface. (Usually the internal interfaces within a network. See UNI.)
NPR	Noise Power Ratio.
NRZ	Non Return to Zero. (A binary encoding technique for transmission of data.)
NRZI	Non Return to Zero Inverted. (A binary encoding techniqe for transmission of data.)
NSAP	Network Service Access Point. (Prime address point used within OSI.)
NT	Network Termination. (Termination designed within ISDN e.g. NT1 and NT2.)
NTE	Network Terminating Equipment. (Usually refers to the customer termination for an ISDN line.)
NTU	Network Terminating Unit. (Used to terminate subscriber leased line.)
OA&M	Operations, Administration and Maintenance. (Also written as OAM.)

OAM&P	Operations, Administration, Maintenance & Provisioning.
O&AM	Object & Attribute Management. (NM Forum.)
O&M	Operations & Maintenance.
OC	Optical Carrier. (SONET standard for optical signals, e.g. OC-1 at 51.84Mbit/s.)
OCR	Optical Character Recognition. (System which can read characters by optical scanning and converting these into electrical signals for processing.)
OEM	Original Equipment Manufacturer. Supplier who makes equipment for sale by a third party. The equipment is usually disguised by the third party with his own labels.)
OKQPSK	Offset Keyed Quaternary Phase Shift Keying.
OLR	Overall Loudness Rating. (Measurement of end to end connection for transmission planning.)
OLTU	Optical Line Terminating Unit. (Or Optical Line Terminal Unit. Equipment which terminates an optical line, usually converting optical signals to electrical and vice versa.)
ONT	Optical Network Termination. (Termination point for optical fibre access system.)
OOK	On-Off Keying. (Digital modulation technique. Also known as ASK or Amplitude Shift Keying.)
OTDR	Optical Time Domain Reflectometer.
PAM	Pulse Amplitude Modulation. (An analogue modulation technique.)
PC	Private Circuit.
PCM	Pulse Code Modulation. (Transmission technique for digital signals.)
PDH	Plesiochronous Digital Hierarchy. (Plesiochronous transmission standard.)
PDM	Pulse Duration Modulation. (Signal modulation technique, also known as Pulse Width Modulation or PWM.)

PDU	Protocol Data Unit. (Data and control information passed between layers in the OSI Seven Layer model.)
PEP	Peak Envelope Power.
PF	Power Flux Density. (Measure of spectral emission strength.)
PFM	Pulse Frequency Modulation. (An analogue modulation technique.)
PLL	Phase Locked Loop. (Technique for recovering the clock in transmitted data. Often performed by an integrated circuit.)
PM	Phase Modulation. (Analogue signal modulation technique.)
POH	Path Overhead. (Information used in SDH transmission structures.)
PON	Passive Optical Network. (Technology for implementing fibre optic cable access in the local loop.)
PPM	Pulse Phase Modulation. (Analogue modulation technique often called Pulse Position Modulation.)
PRBS	Pseudo Random Binary Sequence. (Signal used for telecommunication system testing.)
PRF	Pulse Repetition Frequency. (Of a pulse train.)
PRK	Phase Reversal Keying. (A modification to the PSK modulation technique.)
PSK	Phase Shift Keying. (Analogue phase modulation technique.)
PSN	Packet Switched Network.
PSN	Public Switched Network.
PSS	Packet Switched Service. (Data service offered by BT.)
PSTN	Public Switched Telephone Network. (Term used to describe the public dial up voice telephone network, operated by a PTT.)
PTN	Public Telecommunications Network.
PTO	Public Telecommunication Operator. (A licensed telecommunication operator. Usually used to refer to a PTT.)

| PTT | Postal, Telegraph and Telephone. (Usually refers to the telephone authority within a country, often a publicly owned body. The term is also loosely used to describe any large telecomunications carrier.) |
| PWM | Pulse Width Modulation. (Analogue modulation technique in which the width of pulses is varied. Also called Pulse Duration Modulation, PDM, or Pulse Length Modulation, PLM.) |

QAM	Quadrature Amplitude Modulation. (A modulation technique which varies the amplitude of the signal. Used in dial up modems. Also known as Quadrature Sideband Amplitude Modulation or QSAM.)
QD	Quantising Distortion.
QoS	Quality of Service. (Measure of service performance as perceived by the user.)
QPRS	Quadrature Partial Response System. (Signal modulation technique.)
QPSK	Quadrature Phase Shift Keying. (Signal modulation technique.)

RBER	Residual Bit Error Ratio. (Measure of transmission quality. ITU-T Rec. 594-1.)
RBOC	Regional Bell Operating Company. (US local carriers formed after the divestiture of AT&T.)
RF	Radio Frequency. (Signal.)
RFI	Radio Frequency Interference.
RIN	Relative Intensity Noise. (Measure of noise in an optical source.)
RPFD	Received Power Flux Density.
RTS	Request To Send. (Handshaking routine used in analogue transmission, such as by modems.)

| SDH | Synchronous Digital Hierarchy. |
| SEDC | Single Error Detecting Code. (Transmission code used for detecting errors by use of single parity checks.) |

SELV	Safety Extra Low Voltage circuit. (A circuit which is protected from hazardous voltages.)
SES	Severely Errored Second. (Any second with a BER exceeding 10^{-3}, as per ITU-T G.821.)
SHF	Super High Frequency.
SIO	Scientific and Industrial Organisation.
SMX	Synchronous Multiplexer.
SNI	Subscriber Network Interface.
SNR	Signal to Noise Ratio.
SONET	Synchronous Optical Network. (Synchronous optical transmission system developed in North America, and which has been developed by ITU-T into SDH.)
SQNR	Signal Quantisation Noise Ratio.
SSB	Single Sideband.
SSBSC	Single Sideband Suppressed Carrier modulation. (A method for amplitude modulation of a signal.)
STDM	Synchronous Time Division Multiplexing.
STDMX	Statistical Time Division Multiplexing.
STE	Signalling Terminal Equipment.
STM	Synchronous Transport Module. (Basic carrier module used within SDH, e.g. STM-1, STM-4 and STM-16.)
STMR	Sidetone Masking Rating. (Measure of talker effects of sidetone.)
STP	Shielded Twisted Pair. (Cable.)
STP	Signal Transfer Point.
STS	Space-Time-Space. (Digital switching method.)
STS	Synchronous Transport Signal. (SONET standard for electrical signals, e.g. STS-1 at 51.84Mbit/s.)
STX	Synchronous Transmission crossconnect. (Crossconnect used within SHD.)
STX	Start of Text. (Control character used to indicate the start of data transmission. It is completed by a End of Text character, or ETX.)
TA	Telecommunication Authority.

TCM	Time Compression Multiplexing. (Technique which separates the two directions of transmission in time.)
TCM/DPSK	Trellis Coded Modulation/Differential Phase Shift Keying.
TCP/IP	Transmission Control Protocol/Internet Protocol. (Widely used transmission protocol, originating from the US ARPA defence project.)
TDD	Time Division Duplex transmission.
TDD/FDMA	Time Division Duplex/Frequency Division Multiple Access. (means of multiplexing several two way calls using many frequencies, with a single two way call per frequency.)
TDD/TDMA	Time Division Duplex/Time Division Multiple Access. (Means of multiplexing two way calls using a single frequency for each call and multiple time slots.)
TDM	Time Division Multiplexing. (Technique for combining, by interleaving, several channels of data onto a common channel. The equipment which does this is called a Time Division Multiplexer.)
TDMA	Time Division Multiple Access. (A multiplexing technique where users gain access to a common channel on a time allocation basis. Commonly used in satellite systems, where several earth stations have total use of the transponder's power and bandwidth for a short period, and transmit in bursts of data.)
TE	Terminal Equipment.
TELR	Talker Echo Loudness Rating. (Overall loudness rating of the talker echo path.)
TM	Trade Mark.
TMA	Telecommunication Managers' Association. (UK)
TMN	Telecommunications Management Network.
TNV	Telecommunication Network Voltage circuit. (Test circuit for definition of safety in telecommunication systems.)

TPON	Telephony over Passive Optical Networks.
TTE	Telecommunication Terminal Equipment.
TTY	Teletypewriter. (Usually refers to the transmission from a teletypewriter, which is asynchronous ASCII coded.)
TeraFLOP	(Trillion Floating Point Operations per second. (Measure of a super computer.)
UAP	User Application Process.
UART	Universal Asynchronous Receiver/Transmitter. (The device, usually an integrated cirucit, for transmission of asynchronous data. See also USRT and USART.)
UDF	Unshielded twisted pair Development Forum. (Association of suppliers promoting transmission over UTP.)
UHF	Ultra High Frequency. (Radio frequency, extending from about 300MHz to 3GHz.)
UI	User Interface.
UL	Underwriters Laboratories. (Independent USA organisation involved in standards and certification.)
UNI	User Network Interface. (Also called User Node Interface. External interface of a network.)
UPS	Uninterrupted Power Supply. (Used where loss of power, even for a short time, cannot be tolerated.)
USART	Universal Synchronous/Asynchronous Receiver/Transmitter. (A device, usually an integrated circuit, used in data communication devices, for conversion of data from parallel to serial form for transmission.)
USB	Upper Sideband.
USRT	Universal Synchronous Receiver/Transmitter. (A device, usually an integrated circuit, which converts data for transmission over a synchronous channel.)
UTP	Unshielded Twisted Pair. (Cable.)
VDU	Visual Display Unit. (Usually a computer screen.)

VFCT	Voice Frequency Carrier Telegraph.
VHF	Very High Frequency. (Radio frequency in the range of about 30MHz and 300MHz.)
VHSIC	Very High Speed Integrated Circuit.
VLF	Very Low Frequency. (Radio frequency in the range of about 3kHz to 30kHz.)
VLSI	Very Large Scale Integration. (A complex integrated circuit.)
VNL	Via Net Loss. (The method use to assign minimum loss in telephone lines in order to control echo and singing.)
VQ	Vector Quantisation. (Encoding method.)
VQL	Variable Quantising Level. (Speech encoding method for transmission of speech at 32kbit/s.)
VRC	Vertical Redundancy Check. (Parity method used on transmitted data for error checking.)
VSB	Vestigial Sideband modulation. (A method for amplitude modulation of a signal.)
WACK	Wait Acknowledgement. (Control signal returned by receiver to indicate to the sender that it is temporarily unable to accept any more data.)
WDM	Wavelength Division Multiplexing. (Multiplexing technique used with optical communications systems.)
WDMA	Wavelength Division Multiple Access. (Multiple access technique.)
WIMP	Windows, Icons, Mouse and Pointer. (Display and manipulation technique for graphical interfaces, e.g. as used for network management.)

Index